THE OLD DATA MINER

ABRAHAM BOYARSKY

Paperback ISBN: 979-8-9886354-2-0
E-book ASIN: B0CP4F2QND

Printed in the United States by Bayou Wolf Press
Bayou Wolf Press
Mobile, Alabama
USA
www.bayouwolfpress.com

❀ Created with Vellum

Dedication - TBF

PRAISE FOR ABRAHAM BOYARSKY

A unique and exceptional voyage that fuses the mystical world of Kabala with mathematics, biology and computer science... At a time when faith and science increasingly collide in public conversation Boyarsky's fascinating novel presents us with both the boundless energy of human imagination and the limits and consequences potentially entailed by daring to approach the realm of hidden divine knowledge.

JOEL MILLER - MONTREAL FILMMAKER

CHAPTER
ONE

MONTREAL, Wednesday March 11, 2020.

Minutes after the nightly curfew ended, Jacob Lazerson left his apartment and stepped into the ankle-deep snow. Two hours later, on returning home from morning services in the synagogue, his socks were soaked and the cold had numbed his toes. Melted snow fell from the brim of his hat into his long white beard. Absorbed in thoughts about his computer experiment, the 80-year-old retired mathematician scarcely noticed the late winter storm.

Often, away from his desk, he would mull over the verse in the Zohar that had inspired what would be his final project: "God looked into the Bible and created the universe," which the sages elaborated upon through a parable of a king of flesh and blood who built a castle with the help of an architect. The architect, in turn, designed it based on the wisdom gleaned from scrolls and books. Similarly, God gazed into the Bible and created the universe.

Jacob never doubted that the Bible served as the blueprint for creation, believing that within its pages lay the very DNA codes of life. Yet faith alone wasn't sufficient for this zealous scientist, determined to unravel the divine process by which God transformed Hebrew text into DNA sequences, thereby crafting a world with myriad life forms. Armed with a long strand of DNA, a digital version of the Five Books of Moses, and a Java program he himself had coded, Jacob set out on the most ambitious research endeavor of his lifetime: utilizing traditional scientific methods to validate the divinity of the Bible and consequently bring about the Messianic Age. So firm were his convictions, that he had already prepared a draft for the Proceedings of the National Academy of Sciences, bearing the bold title, "A Computer-Aided Proof for the Existence of God: A Scientific Approach to Summon the Messiah."

Trudging through the snow, Jacob approached a building—one in a series of identical red brick three-story structures built in the 1940s—where he lived with his wife, Sarah. His thoughts wandered to the family's savings he had lost in the dot com crash two decades prior. Their spacious home on a tranquil, tree-canopied avenue in the center of the Jewish community had been sold in haste to address looming margin calls. A few months later, they had to sell their country cottage, where he had spent many happy summers with Sarah and their daughter, Elka. Now confined to a humble flat in this bug-ridden building, their survival hinged on government pensions. No words from Sarah could ease the weight of guilt that Jacob bore for plunging them into such impoverishment.

In the small lobby, a stench from broken sewer pipes greeted Jacob. Drain flies skittered on the walls and the interior door's glass. Brushing his left forearm against the stairwell wall, he ascended the steps with deliberate slowness, pausing for breath at each landing. The intense labor he had poured into his algorithm over the recent months had weakened him considerably.

Standing at his front door, Jacob imagined himself tweaking one of the universal physical constants: the speed of light in a vacuum or the Planck constant, then observed the universe devolve into its primordial state. He looked around and sighed with relief that the world was in tact as before his interference and thanked God for His immutable laws through which nature testifies to His faithfulness. Jacob reverently touched the mezuzah fastened to his apartment's doorframe, turned the key, and stepped inside.

"How's my *maydelle* this morning?" Jacob quipped as he approached Sarah in the tiny vestibule, water droplets falling from his coat onto the aged red carpet. The tender Yiddish word for 'little girl' always brought forth a smile to Sarah's face, deepening the crescent wrinkles surrounding her mouth.

"Thank God," she responded. Just moments before Jacob's entry, a white mouse had darted across the kitchen floor. Sarah kept this to herself, fearing he might interpret it as a complaint. She stood swathed in a baggy black sweater, its sleeves frayed. A kerchief concealed most of her hair, but wisps of silver curled about her ears, reminiscent of a young boy's side locks. Taking the prayer sack from Jacob's hand, she then hung his coat on a nail positioned between the front door and the closet. The black ball point splotch at the bottom of Jacob's white shirt pocket drew her attention. She had tried many times to scrub off the stain that looked like a bullet wound in the heart.

Jacob settled into his chair, removed his shoes, then pulled off his socks before sliding his feet into crocs that were spattered with toothpaste. Full of avid anticipation, he made his way to his tiny office nestled in a corner of the living room. Papers littered the desk—a collection of holy books on the front left and a stack of mathematics books on the front right. Anytime he found a mathematics book erroneously atop a holy one, he berated himself for the oversight. Hunching over his desk, he nudged the mouse with his right hand, impatiently waiting for the computer screen to come alive. A twinge of pain shot through the joint of his curved right forefinger, the result of years spent pressing chalk to blackboards.

Jacob rubbed his eyebrows, releasing a shower of dandruff that drifted leisurely downward in a Brownian motion. On the left side of the computer screen, a window showed that the CPU was diligently processing datasets of Hebrew texts and DNA sequences. The ongoing operation of the program indicated that the moment he was fervently waiting for had still not been attained. He scanned a column, searching for the number 0 under the variable labelled HD which measures the Hamming Distance between two sequences of identical length. The number 0 implies a perfect match between sequences.

After examination of the data on the screen, Jacob turned his attention to news websites that covered the rapidly proliferating corona virus. A headline from The New York Times proclaimed, 'Plague on a Biblical Scale: Chassidic Families Severely Afflicted by Virus.' With hospitals gearing up for an influx of patients, mandatory quarantines were strictly enforced. However, many Jewish communities remained defiant, disregarding the directives to quarantine, wear protective masks outside their homes, and maintain social distancing. Weddings and large parties went on as usual, casting a grim shadow over the elderly demographic. In Montreal, hundreds of guests from Brooklyn arrived to celebrate a wedding of prominent families. The visitors danced and rejoiced with their relatives and friends, leaving behind a generous dose of the corona virus, unaware that they were seeding a path of infections in their wake. Jacob encapsulated these tumultuous events with a metaphorical image of a serpent stealthily manoeuvring through a dense jungle, hissing ominously in the depths of the night, until it unleashed its venom on humanity in the winter of 2020.

Sarah's voice beckoning him for breakfast, broke his train of thought. Jacob entered the kitchen, noting the worn black and white vinyl tiles that revealed patches of the grimy wood underneath. Years had yellowed the stucco walls, and rainwater leaking from the roof had damaged the ceiling.

"Does it still hurt?" Sarah's voice carried a note of concern, referring to the bruise on his forehead. Late one night the previous week, Jacob had gone to the bathroom. As he lifted a cup to rinse his mouth, it slipped out of his wet hand. On bending over to retrieve it, his forehead struck the rim of the sink.

"I wish it had been me," she said, as always a partner in his pain and tribulations.

"God forbid!" Jacob replied instantly, recognizing the sincerity in her words. Sarah drew his head to her lips, lovingly kissed the swollen purple contusion, then rubbed it with her fingertips to diffuse the congealed blood beneath the skin.

"This project is draining you," she observed, her tone suddenly laden with sorrow. She had never witnessed him so consumed by a task. "I feel so helpless. Is there any way I can help you?" she implored, a trace of hope in her voice.

"You'll be able to soon," he responded, albeit cryptically.

A blue milk crate was positioned on the kitchen table, its contents visible. It housed the holy books that Sarah frequently referred to for her daily studies. Each book showed signs of extensive use, especially near the bottom frayed right corners where her fingers, moistened by her tongue, flipped the pages.

Removing a number of pennies from a nearby plastic container, Jacob dropped them into a charity can situated beside their old Telefunken radio. Gone was their china cabinet that had once proudly displayed their expensive dishes and vases. The only remnant of their bygone affluence was a silver tureen, its ornate design catching the light. Standing atop the refrigerator on its alligator feet, the tureen held sentimental value for Sarah. A cherished wedding gift from her mother, Sarah meticulously cleaned and polished it before every Sabbath. Engrossed in this routine, she would often lose track of time, only to be gently reminded by Jacob that the Sabbath was fast approaching. Behind the tureen was a small *shofar* that Jacob blew on the High Holidays for neighbours unable to attend the synagogue.

Jacob sat down at the narrow table, snot suspended from his nose. Sarah handed him a paper towel, then set a bowl of porridge before him, next to his medications. Porridge had become their staple diet, except on the Sabbath when cooking was forbidden. Sink water cooled the thick hot mixture. Sarah served Jacob coffee in the only cup that still had a handle. All the other cups had fallen from her hands over the years. A large urn of perpetually boiling water was the one luxury in this austere household. In their former home, folded cloth napkins and stainless steel cutlery adorned the breakfast table; now they were replaced by pieces of toilet paper and plastic cutlery that was reused. Jacob's visible anxiety, apparent in the nervous shaking of his left leg, prompted Sarah to place her hand gently on his knee, bringing it to an abrupt stop.

"Please try to relax," she entreated.

"Sorry," he whispered, his eyes lowered.

"For what?" she queried.

"For all the suffering I've brought upon you."

"I don't know what you're talking about," Sarah retorted,

dismissing his words with her characteristic wave of the right hand. "Did we ever vacation in Miami when we had money?"

"No, but the option was available if we had wanted," he remarked, a hint of illusory nostalgia in his voice.

"You know very well we would never have travelled anywhere." Cradling her cup of hot water, Sarah attempted to thaw her icy fingers.

"Your porridge is cooling," she gently reminded him.

He recited the customary blessing for grains and consumed a few spoonfuls. Sarah, in her usual manner, set her reading glasses aside while eating. Jacob took the initiative to clean them for her, employing his breath and the edge of the tablecloth. Noticing a tiny speck of porridge on her face, he gently wiped it away with his finger. He then pushed his chair back to stand up, eager to start the day's work.

"You've hardly eaten anything. The computer can wait," Sarah asserted, emphasizing her concern over his obsessive dedication to the algorithm. He acquiesced with a slight nod. The faint melody of a piano was heard from the floor below, evoking memories of Sarah's piano sessions in their previous home. He fondly recalled how her renditions had provided a serene backdrop to his mathematical reflections. The sale of their house had included the piano, a condition set by the buyer. Despite the loss, Sarah had transitioned without resentment from her cherished piano to the Book of Psalms. Soon after their relocation, she had embarked on giving piano lessons to children within the community. However, her inherent kindness was often exploited; many ceased payments but Sarah, true to her nature, continued giving lessons as if she was getting paid. When the pandemic struck, she was asked not to come anymore. After retirement, Jacob tried tutoring mathematics at a nearby high school. A brief stint made him realize he was not cut out for this.

Jacob expressed his gratitude for the meal, then went to his desk. Turning to his left side, he could see the kitchen clock. It chimed every hour, a regular reminder that humanity was one hour closer to the Messianic Age. Dominating the space between the window behind Jacob's desk and the balcony door was a prominent portrait of the Rebbe. As the spiritual leader of the Chassidic dynasty Jacob belonged to, the Rebbe's image held significant importance. A friend had snapped the photograph and gave it to Jacob. Sarah had it blown up and mounted on a wooden frame. During the move to their apartment, the protective glass had shattered. Jacob had removed the portrait from the frame and taped it directly to the wall. Below it was a brown photograph from 1941, capturing Jacob's mother with her infant Jacob at the main Leningrad train station. Jacob's right hand rested on his mother's right forearm, while his left hand was raised, straining backwards to touch her, to affirm her presence. The woman's face was averted modestly from the camera held by Jacob's father, who later starved to death during the siege of Leningrad.

Through gaps between floral swirls of ice on the window, Jacob saw snow flakes floating into the crown of a poplar between apartment buildings. From the top of the exterior window frame, icicles tapered downward like a colonnade of tornadoes. A speck of ice bonded a yellow autumn leaf to the corner of the window sill where a loyal seagull with dirty white plumage made its home. Every morning Jacob left bread crumbs on the porch floor. Occasionally the seagull would gaze inside with its splayed eyes, as if wondering what the old man behind the computer monitor was up to. Across the alleyway, South Asian children peered through their windows, their attention drawn to Jacob. One of them held a console in his hands. Jacob imagined himself as a character in the boy's video game. He raised his hand to scratch his nose and wondered whimsically if it was his decision to do so or his response to a wireless command sent by the child.

Warm summer evenings came to mind, when Sarah sat on the back porch singing Chassidic songs and recounting Biblical stories to the children on their back porches.

"Who is Moses? Who is God? Where does God live?" they asked inquisitively. Sarah explained the story of creation and the miraculous history of the Jews. As the children grew older, she taught them the 7 rational precepts which all humans are commanded to observe.

On the cover of a coffee stained manila folder, a few handwritten sentences outlined Jacob's strategy to search for a sequence of Hebrew letters in the Bible that matched exactly the letters in an exon, a DNA sequence that codes for a protein. An exact match for a sequence of 10 numbers is as unlikely as winning the Mega Lottery and the probability of precisely matching a thousand numbers is unimaginably smaller. Yet it was such a statistical anomaly that Jacob sought from his algorithm. When he finished jotting down the computational details of his plan and contemplated them deeply, he saw clearly the final outcome in his mind's eye. If his search algorithm could identify the associated Hebrew sequences for all exons in the human genome, it would prove beyond a doubt that the Bible was written with divine intelligence.

Jacob discovered a lengthy exon he named X, written in the language of the four DNA bases: adenine (A), cytosine (C), guanine (G), and thymine (T). At first glance, X seemed like a senseless sequence, GGAGAGCAGGATACCACAGCCTG......., stretching over 959 letters. Assigning a unique height to each of the four letters, Jacob plotted their heights consecutively along a line, seeking recurring patterns and symmetries on this genetic terrain. However, the sequence remained elusive, its intricate design masking within it an ostensible order. Despite this chaos, the precision must be perfect; a single mutation could alter its corresponding protein and potentially trigger cancer.

To compare DNA sequences with sequences from Biblical texts, Jacob's initial task was to categorize the 22 Hebrew letters into four groups. These groups should reflect the same probability of appearing in a 959-letter Hebrew sequence as A, C, G, T did in exon X. This posed a bin packing problem: how to align the occurrence probabilities of the 22 Hebrew letters with the probabilities of the four DNA letters in X. Given the exceptional length of exon X, Jacob recognized the challenge of identifying a perfect Biblical match. However, such a discovery would have profound scientific implications.

Jacob embarked on his project with a great deal of confidence. He had hoped to find the perfect match to exon X in Genesis, even among the first few chapters. But he soon realized that God was more subtle than he could have ever imagined. There were enough combinations for most of the exons in Genesis. But when he searched the internet, he found exons longer than the number of Hebrew letters in Genesis and was forced to conclude that since there are 5 books in the Bible, 5 books are needed for creation.

Before commencing his coding, Jacob wanted to ensure that the Bible's complete text, consisting of 304,805 Hebrew letters, was adequate to match the length of the longest human exon. A Google search unveiled that the human exon TTN, responsible for the titin protein, consists of 304,814 DNA letters and is situated on chromosome 2's elongated arm. Jacob felt reassured when he discerned the near-identical letter count between the Hebrew Bible and the TTN exon.

Following the coding of the packing algorithm, Jacob wrote a program to assess the similarity between a DNA sequence and a Hebrew letter sequence of equivalent length. When the packing problem had multiple solutions, it occasionally demanded hours to calculate every possible combination, subsequently displaying the smallest HD between X and all Hebrew sequences of identical length. After completing one comparison, the Hebrew sequence was shifted one letter to the left. This procedure was repeated 304,856 times, until all possible 959-letter sequences in the Bible were compared with exon X in the HD metric. Jacob dubbed this method, "sweeping to the left across the Bible."

Upon finalizing and verifying the algorithm, Jacob asserted with a sense of historical reckoning: "Zyklon B has been supplanted by exon X." He visualized the results of his great experiment as if he was admiring a masterpiece by Rembrandt.

CHAPTER
TWO

THE SOUND of the seagull flapping its wings against the window pane distracted Jacob from his work. In a genial mood, he turned to address the bird: "Dear Mr. Seagull, how are you today? I'm blessed to have made your acquaintance. You stand kind and nonjudgmental behind the glass. I value that and promise a reward soon. The moment the Hamming Distance hits 0, you'll be the first to know. Like Noah and the dove, I'll send you off to announce to the world that the Messiah has finally arrived!"

Although the algorithm had processed over 90% of the Biblical text without success, Jacob's expectations were not diminished. There was still time for the scientific method to deliver the illustrious result, perhaps in a mere instant. The word instant evoked a saying from the Rebbe, "The advent of the Messianic age is only a gesture away."

"But which gesture, by whom, and when?" Jacob pondered. "Perhaps just a tap on my keyboard, by me, and now?"

Mid-morning, Jacob felt Sarah's presence beside him. Stooping slightly, she held a letter she had written to the Rebbe, intending for Jacob to fax it to the office near the Rebbe's gravesite in New York.

"Elka asked me to write for a friend who caught the virus," she explained. "I hope Elka stays safe. Working in a kindergarten is dangerous. I've also asked for blessings for Elka, the children, you, and the success of your program."

"Thank you for including the program," he replied.

"If it matters to you, it matters to me."

"It's significant not just for me, but for all of humanity," he affirmed, a intimation of pride in his voice.

"I know, but I wish you would take care of yourself like everyone else," she remarked regretfully. Adjusting his askew black skull cap, she reminisced about how he had worn that same cap during his university days, stained with chalk marks that never washed off. Jacob was healthier during his teaching days. He had maintained a regular routine, ensuring he was fresh for morning lectures. But a heated dispute with the department over a student's exam had confirmed to Jacob and Sarah that it was time for him to retire.

"The program is nearing its end," Jacob reassured her. "Another week or two, at most. Once I prove the concept with exon X, I'll adapt the algorithm for exons associated with cancer and other ailments. I even have an idea for the corona virus. Exons with mutations might closely match specific sequences in the Bible. This could locate the mutations and suggest modifications." Eagerly he continued, "The corona virus has a region called the RBD that allows it to bind with human ACE2 cells. There's research on how mutations affect this binding. With my algorithm, I might trace these mutated proteins back to their Biblical sequence. Maybe, with divine intervention, I could design a vaccine based on this mutation. It's just a thought for now."

Accepting his zeal, Sarah responded, "I'll recite Psalms for your program's success."

As she moved away, Jacob envisioned her surrounded by a protective halo of Psalms. As he listened to her melodious hum, he could not help but pause his work and hitch a ride on her holy words. He reflected on the subtle race between them: who would bring the Messiah first - she with her Psalms or he with his algorithm? It was a sprint from opposite directions to the peak of the mystical Mount Everest.

The soles of Jacob's feet tingled when he did not move them for long periods. He stood up to stretch his legs. Through the window,

Jacob heard the naked branches, sheathed in frozen myelin, clanging musically. The wind had blown away most of the snow from the tree, but some remained, balls of cotton in the forks of branches. Jacob's gaze turned to the open space between apartment buildings along Cote St. Catherine which afforded a narrow view of McKenzie King Park and, beyond the park, the playground of a Jewish school. An Israeli flag in the school yard fluttered in the wind. How ironic, he thought, that the school stood facing a park named after a prime minister who had prohibited a single Jew from entering Canada during the Second World War.

Returning to his desk, Jacob was startled by a live white mouse sitting on top of his white computer mouse, its tail in perfect alignment with the wire. A clap of his hands sent the mouse leaping from the desk to the floor, then to the kitchen. Jacob marveled at the strange incident that had virtually zero probability of occurring. But the incident worried him: once his computer experiment was proven successful, skeptics may argue that if a live mouse can align itself perfectly with a computer mouse, what was so remarkable about aligning a DNA sequence X with a 959 Hebrew sequence? But Jacob was able to dispel his concern by reasoning that the algorithm would be capable of finding Hebrew sequences for all exons, whereas thousands of white mice aligning perfectly with thousands of white computer mice had no basis in logic or probability theory.

A pop-up on his screen from a Jewish website announced a corona virus death in Crown Heights, Brooklyn: Rafael Siminovitch, aged 81, whom Jacob had known since their childhood days in Samarkand. While Jacob and his mother had managed to escape Russia with forged Polish passports, Rafael and his family were left behind. Rafael was caught teaching children Judaic subjects in an underground classroom. For that crime he served 14 years in the Gulag. In 1971 he arrived in America, a physically broken man.

Jacob recalled the classrooms of his childhood: dank and dark cellars. In one of those rooms, a 6 year old boy innocently questioned his teacher about the existence of God. The teacher wrapped his hand in a handkerchief and waved it back and forth.

"What is moving?" he asked.

"The handkerchief," a few of the boys answered.

"What is making the handkerchief move?"

"Your hand!" the boys shouted.

"But do you see my hand?" the teacher prompted.

"No."

"Does my hand exist?"

"Yes," many of the boys answered, but not Jacob.

"You don't see my hand but it exists," the teacher asserted as he revealed his hand. All the children, except Jacob, were convinced of God's existence. Even now, in old age, Jacob was attempting to peel the handkerchief from the Hand of God.

In the early afternoon, Jacob's eyes drooped. His face sunk into his beard and he dozed off. He dreamt about the DP camps of his childhood after the war, wandering among the ruins. The children called him *shpilkes*, the Polish word for safety pins, because his ragged clothes were held together by dozens of *shpilkes*. The Jewish men spoke among themselves in Russian hoping the children would not understand. But Jacob did understand and overheard an adage that never left him: 'nothing comes easy except peeing in the rain.'

Jacob had not moved from his chair for six hours. When he tried to stand up, his back ached. It was time for the afternoon service. Sarah had left dry socks on the back of a kitchen chair. After Jacob drew them up over his cold feet, he pulled up his boots and prepared to go to the synagogue. Sarah helped him into his coat that was still damp from the morning snow. "Please wear a mask even if no one else does," she cautioned him as she handed him a blue mask and a list of grocery items to buy on the way home. Beneath his coat in a shopping bag, he had hidden Sarah's tureen. After leaving the apartment building, he took the bag out and carried it in his hand. The streets were slippery. Jacob walked through Van Horne Park as the lambent shadows of nightfall spread silently toward him. He was thinking about the survivors of the war who used to frequent this park in the sixties and seventies. Jacob had made friends with many of them. Hershel Sontag, who had jumped off a train heading to Mauthausen, had witnessed the murder of his 3 children. On a bitterly cold night, he had crossed a frozen river on wooden planks. Behind him, a man fell through the ice. Hershel had tried in vain to save him. That scene now resided only in the memory of one old man who was plodding through the slush of a dark, deserted park. Glancing over the duplexes along Van Horne Avenue, Jacob saw the light on the wing tip of an airplane flickering slowly in the gloaming.

The weekday synagogue was located in the basement of the community boys school. Jacob hung up his coat in the cloakroom, then proceeded into the chapel. He was pleased that no one here knew of his clandestine existence as a manipulator of symbols and numbers. To his surprise, several men were wearing masks. Hands extended for greetings were playfully swapped at the last second with impromptu elbow taps. Elderly men, distinguished by patriarchal beards and black fedoras, sat hunched over sacred texts, the hair on their napes like the sparse feathers on a chicken's pink skin.

Many of Jacob's contemporaries had ascended to a higher realm, now replaced by young men with black beards, engrossed in conversations on their smartphones. Joseph felt more at ease in the back of the chapel. An albino man stood nearby, squinting into a prayer book he held inches from his face. Jacob had never encountered him but felt a bond. Both were outsiders here: the albino due to his stark appearance, and Jacob because of his unconventional pursuit to summon the Messiah using scientific methods.

He learned that the petite, aged Chassid, Yechiel Horowitz, had passed away from the corona virus earlier that day. Before the war, Yechiel had studied in Poland, and until his last breath, remained a staunch believer in the imminent arrival of the Messiah. Legend said that if Yechiel were to bleed, it would be the Messiah's blood that would flow. Jacob fondly recalled how Yechiel would enter the synagogue, a glint in his eyes, seeking a pair of glasses. Several were always strewn about on tables, but whichever pair Yechiel chose, amazingly corrected his deteriorating vision. With a contagious chuckle, he would habitually adjust fellow congregants' jacket collars, whether they needed the adjustment or not. It was a gesture of love. Jacob had been confident that Yechiel would live to welcome the

Messiah. Now Jacob wondered who among them would merit to witness the advent of the Messianic era.

A tall young man entered the synagogue, paused, looked left, right, then straight ahead. Leaning slightly forward like a skier at the top of a slalom, he sashayed into the synagogue. Near the Ark along the western wall, the languid-faced Solomon stood up. His periscopic neck swivelled slowly back and forth across the busy chapel, then he sat down. Nearby, a quorum of 10 men had begun reciting the Amidah prayer, the core afternoon service. Jacob glanced around at the black hats and saw a field of sunflowers swaying synchronously in a gentle breeze. He picked up a prayer book from a table and joined them.

After the service, Jacob sat down on a bench opposite the elderly Rabbi Cohen. Once an illustrious Torah savant, he now stared at a wall adorned with copper plaques memorializing former congregants, his vacant eyes suggesting the onset of advanced dementia. A friend attempted to draw him into conversation, but upon receiving no response, raised his voice in exasperation: "Wake up already! Wake up!"

In the interim between the afternoon and evening services, many men chanted Psalms, praying for a local teenager battling the corona virus in an intensive care unit. Jacob added his voice to the lengthy chapter 119. Meanwhile an elderly Yemenite man was collecting donations for a toddler diagnosed with leukemia. Jacob handed him two loonies. Without examining the donation, the man expressed deep gratitude, as though the sum was substantial.

Shmuel Shteiner, affectionately called Shmuki, entered. Though only 5'4", his posture and expanded chest gave the illusion of greater height. Spotting Jacob, Shmuki approached, playfully grasping Jacob's long beard and feigned yanking himself up as if ascending a fire pole.

"Remove that silly thing," he teased, motioning to Jacob's mask. Hidden beneath the mask, Jacob's lips curled into a smile.

"Sarah made me promise," Jacob countered.

Their friendship dated back to their adolescent years. Later, while Jacob pursued university studies, Shmuki remained in the yeshiva environment, emerging as the Rebbe's informal emissary in Montreal. Even after Jacob's marriage, Shmuki often included the more reserved man in his fervent ventures to spread Torah observance among Jews. Although Shmuki never married, he exuded joy, revelling in the family celebrations of others as if they were his own.

Following the evening prayers, a group of men gathered in a circle around the podium. With hands resting on each other's shoulders, they danced with jubilation. Their voices rose in unison, passionately chanting, "We want the Messiah now! We don't want to wait!" On the podium, children sang with fervour, their hands spiritedly marking the rhythm. The chant morphed into a new song, "Long Live our Master, our Teacher, and our Rabbi, King Messiah, for all eternity."

As Jacob discreetly exited the synagogue, he could not help but think, 'If only these men knew how close I am to ushering in the Messianic Age.'

Then his thoughts carried him further and he contemplated what he would do when the algorithm will have matched exon X with a Hebrew sequence. Once the divine power of the Bible was established, the next step would be to search for answers to life's problems in its verses. This implied the development of a search engine, a Google Bible. His first query would be: "Why was it necessary for millions of children to die in the Holocaust?" Jacob feared the answer and worried for God, wondering if He had the ability to hide from Himself.

CHAPTER
THREE

IN THE GROCERY STORE, Jacob pushed a cart through the turnstile. He was always mindful of his expenses, except when buying lollipops for the children in the synagogue, his first purchase. He then chose a tray-pack of chicken and a frozen loaf of gefilte fish. Whenever he was accompanied by Sarah, she would prevent him from purchasing wheat germ. Remembering how she enjoyed sprinkling it on her porridge during better times, he added a jar to his cart. Next, he took two large bags of porridge.

As he approached the fruit section, the allure of red grapes captivated him. The thought of them in his cereal made him reach out impulsively, but he caught himself quickly, and withdrew his hand as if about to touch red hot coals. He selected three ripe bananas, then proceeded to the vegetable aisle where he filled his cart with two 5-pound bags of potatoes and a small sack of onions. Jacob recalled his mother telling him that during wartime, many people lost their teeth because they could not find onions.

Waiting in line at the cash register, Jacob observed the masked shoppers around him. From the multitude of expressions hidden behind masks, the sorrowful gaze of a woman caught his attention. Silently, he mused, 'I see your eyes from above my mask and try to construct your face from the obscured visual data that passes through my fusiform face area. Then I invoke the Kalman filter in my parietal lobe to calculate the minimum mean square estimate of your features.'

Jacob observed the boy packing the articles into two large shopping bags. Without giving any thought to arranging the items for optimal packing, he just stuffed the articles into the bags regardless of size or shape. 'What a waste of space,' Jacob thought.

Upon exiting the checkout lane, Jacob headed to the front counter to purchase a 6/49 lottery ticket. When Sarah was with him, she always expressed her distaste for gambling. He countered her disapproval with his dreams of the millions they could win to which she replied, "What use have we for such wealth? We lack nothing." He then reminded her of Elka's children who would need money for tuition fees, orthodontic work, and the many charities they could generously support. Unmoved, Sarah said, "Wealth makes itself wings like an eagle that flies off to the heavens," quoting a verse whose source she did not remember.

Carrying two hefty shopping bags, Jacob made his way through the shopping center, heading for Victoria Avenue where the pawn shop was located. As he passed the dollar store where the old Black Derby Delicatessen had been located, he paused to catch his breath and set the bags down. Memories of an incident from the early seventies, involving Shmuki and a riot he nearly instigated, flooded his mind.

A long queue of middle-aged and elderly Jews stood in the cold on the mall walkway, waiting to enter the warm deli that boasted its kosher-style food. A large sign over the entranceway proclaimed: 'Jewish Kosher Style Delicacies.' Jacob and Shmuki took their place in line.

"Shmuki, why are we here?" Jacob asked. "Our presence here might make people think this is a kosher restaurant."

"These people shouldn't be eating non-kosher food!" Shmuki retorted passionately.

Jacob kept quiet as Shmuki leaned towards an older, portly man ahead of them. "Did you hear about the food poisoning incident here last week?" Shmuki asked .

The man, with a darkly tanned face hinting at a recent vacation in the south, frowned in confusion. "I haven't heard anything about it."

"My niece was in the mall when an ambulance came to carry people out," Shmuki continued nonchalantly.

The man looked alarmed and asked: "Are you sure?"

"Absolutely! Would I lie to you?"

Whispers of Shmuki's claim circulated quickly, causing a commotion as the crowd sought more details.

"Unfortunately, ladies and gentlemen, it is true. If you don't believe me, you can call Mrs. Schmeltzer. She lives at 175 Finchley Road in Hampstead. Here's a dime." He reached into his pocket and came up with a coin.

"If that's the case, why are you here?" someone questioned.

Shmuki thrust Jacob forward. "Please meet Professor Erol Shaughnessey from McGill University's Department of Health and Nutrition. He will clarify."

"We're here to alert the community about potential food contamination concerns," Jacob stammered.

At that moment, the manager of the restaurant, a brawny man in his fifties, appeared in the foyer. "Get out of here!" he yelled, waving his fist at Shmuki. "If you're not out of here in ten seconds, I'm calling the cops! One! Two!"

"What a coincidence!" Shmuki countered. "Here are the police!" He whipped out an identification card from his breast pocket, displaying it proudly to the men and women around him. In the top left corner of the card, there was a photograph of Shmuki in a police uniform, cap and all, with a badge on his lapel. "Captain S. Shteiner of the Montreal Urban Community Police Force. At your service!" he announced loudly, saluting smartly, fingers at his temple. The manager stepped forward to examine the identification card.

A distressed woman tugged the manager by the arm, asking urgently, "What about the poisoned food?"

"There is no poisoned food here!" he shouted, defending the establishment

. "Liar!" Shmuki accused, pointing directly at the manager. "I saw you at the funeral of one of your customers! You were wearing a black bow tie and a pink jacket."

"We've been in business twenty-nine years without a single casualty," the manager claimed.

"Liar! Your customers go directly from here to the ER at the Jewish General Hospital. My uncle is the chief doctor there. He told me about the fourteen food poisoning cases from the Black Derby last month. Granted, fatalities have fallen from seven to five this month, but is that a statistic to pride yourself on?"

In a fit of frustration, the manager cried, "He's one of those religious fanatics. He believes we don't serve kosher dishes!" In a sudden move, he knocked off Shmuki's fedora, exposing a black skull cap underneath. "There, you see, I told you!" the manager announced as if Shmuki's skull cap exonerated the restaurant from food poisoning.

Shmuki sensed that the tide of sympathy had turned against him. Now he had nothing to lose.

"Shame on you! Shame!" Shmuki berated them as he and Jacob jostled through the crowd. "Your mothers and fathers were observant Jews. They observed the Sabbath and ate kosher food. But look at you! Look at what's become of you, eating bacon and desecrating the Sabbath!"

"Get out of here!" someone shouted. "Extremists trying to force their crazy laws down our throats!"

Retreating, Shmuki shouted, "Ignorant fools! Idiots!"

Once outside, Shmuki sighed with deep emotion and said, "They need to realize 'kosher-style' doesn't mean kosher." Tears were streaming from his eyes, congealing on his cheeks in the frosty air. Then, all at once, his face loosened into his familiar smile and he said with his usual poise: "Don't worry Jacob. When the Messiah will come, they'll see the truth." In an afterthought, he said: "Then I'll also be able to get rid of the ten thousand pairs of jeans I imported from Taiwan." Jacob glanced quizzically at him. Shmuki elaborated: "When the Messiah will resurrect the dead, they'll need pants. What do you think, they're going to walk around naked? The Messiah will say, 'Shmuki, get them dressed!' And, of course, I'll be ready."

• • •

Jacob lifted the shopping bags and headed toward Victoria Avenue where the small pawn shop was located at the top of an exterior staircase. The shop's front display and walls were crammed with assorted trinkets and electronic devices. Behind a dark curtain, a small space was reserved for passport photo services. Approaching the counter, Jacob hesitantly presented Sarah's tureen to the Sikh man standing there. The man studied the tureen and remarked, "Impressive craftsmanship. I'll give you $250 for it. You've got thirty days to buy it back for $300. After that, it goes on sale."

Nodding, Jacob said, "Understood." He reclaimed the tureen, silently wishing he would never have to resort to such a measure.

After a few minutes on the streets, Jacob was startled by the sudden roar of an airplane flying low in the clouds and instinctively presumed that it was the sound of the ram's horn heralding the arrival of the Messiah. The intensifying noise gave him hope that his algorithm had finally reached its objective.

CHAPTER
FOUR

ON THE SECOND floor of his building, Jacob left one of the
shopping bags at the door of a single mother with a young child,
recent immigrants from Nigeria. He looked around to make sure no
one saw him tie the bag to the door knob. Once in his apartment, Jacob
waited for Sarah to step out of the kitchen, then placed the tureen back
on top of the refrigerator, hoping that she had not noticed its absence.

Having been away from his computer for more than two hours,
Jacob eagerly anticipated the first glimpse at the HD numbers in the
left column on the screen. With a sigh of disappointment at the results,
Jacob returned to the kitchen. Sarah was storing the groceries in the
pantry and the fruits and vegetables in the refrigerator. She arranged
slices of red pepper around a bowl of porridge. With every spoonful,
Jacob took a nibble on the pepper.

After the meal, Jacob sat down on the shabby green sofa along the
wall facing his desk. A book of photographs of the Klondike Gold
Rush had been left behind by the previous tenants. It lay in a corner of
the sofa. The cover showed a long line of men carrying massive packages on their backs, leaning into a steeply inclined snow covered slope.
Jacob saw himself on the Klondike Trail, walking with prospectors past
corpses, men who had died of cold or exhaustion in the brutal Yukon
winter.

Jacob eyed the collection of cardboard boxes scattered around the floor. He reminisced about the early days of his Bible-DNA research, when Sarah's eyes sparkled with the same enthusiasm that had meaningfully filled his retirement. Noticing her genuine intrigue, he had endeavoured to simplify the intricacies of his packing algorithm for her with a tangible example. But first he wanted to explain the Hamming Distance to her. He wrote down the number sequences 10010 and 00011, and showed her that they differ in 2 locations, the first and last positions. Therefore, the HD is 2. The normalized HD number is 2 divided by the length of the sequence, 5. Analogously, the HD between the DNA sequence AACTGTTA and the Hebrew sequence A'C'C'T'T'G'T'A' is 3 since the sequences differ in 3 locations, where A',C',G',T' represent sets of Hebrew letters identified with A,C,G,T by the packing program.

Using cardboard and scotch tape, Jacob had constructed 4 large cubes of different sizes, one side open in each. He also made 22 smaller paper cubes of different sizes. He explained, "The small packages have to fit into the four large boxes in such a way that no space is left over in the large boxes. The packing algorithm finds these perfect fits." He placed the 4 large cubes on one side of the kitchen table and the 22 smaller ones on the other side, then prompted Sarah to pack the smaller cubes perfectly into the 4 large cubes. After trying numerous combinations, she managed to pack one large cube exactly, but failed to pack the 3 other large cubes. At length, she threw up her hands in defeat. Jacob had constructed the 22 small cubes so that a specific combination would solve the packing problem exactly. He had labelled the smaller cubes with the number of the large cube into which they fit. In a matter of seconds, he packed all the small different sized cubes perfectly into the 4 large cubes. There were in fact many other solutions to this packing problem, he told her. Then he explained that the sizes of the large cubes represented the probabilities of occurrence of A,C,G,T in an exon, while the sizes of the smaller cubes represented the probabilities of occurrence of the 22 Hebrew letters in a Biblical text having the same length as the exon. Jacob then told Sarah that the packing program tests millions of different combinations of smaller boxes in a few seconds and prints out only the combinations that form perfect fits in the 4 large boxes. Sarah was amazed.

Shortly after 11 pm Sarah shuffled to Jacob's desk with a cup of coffee and two cookies. She wanted to stay up as late as possible in the hope that this would discourage him from spending another sleepless night. Sarah touched his shoulder, then planted a kiss on his skull cap.

"Please don't stay up late," she whispered imploringly.

"I won't," he replied, but she knew that did not mean much. "Sleep well," he added absently, his back to her, his eyes on the screen where fresh data was streaming in.

Jacob's passionate drive was both a marvel and a concern to Sarah as she watched him absorbed in his work, her heart torn between pride and anxiety. While she admired his dedication and admirable objective, she worried about the toll it was taking on his health.

Jacob had seen photographs of stock exchange floors, men glued to huge screens, frantically following stock prices. Sitting before his screen now, he wondered if he was different than them. Suddenly lightning flashed across the computer screen. Jacob didn't know if it was virtual or really happening in the clouds outside. Opening and clenching his right hand, he tried to relieve the electric pain in his right forefinger. Although he had stopped complaining years ago, his arthritis was not helped by clicking the keyboard all day. Yet his thin nimble fingers were still able to prance across the keyboard with ease.

The tranquil purr of the computer was punctuated by the groans of Jacob's aching joints, but he would not be deterred. The ethereal quest for a connection to what is beyond the tangible world had been ingrained in him since his early days at the yeshiva. There Jacob had learned to sublimate physical acts into spiritual equivalents. He tried to live according to a dictum he had once heard from a teacher: "The horse has to be whipped until it is no longer a horse and flies." This is the divine service of the body, to make it toil in Biblical study until it sucks out the haughtiness concealed in every heart, until the body no longer needs food or rest. Jacob hoped to refine his body by lashing it with hunger and wakefulness until it would become spiritual tweezers with which to grasp a wisp of God's logic and traverse the gap between body and soul.

Sarah brought him a shawl to wrap around his shoulders, trying to provide warmth against the cold that often crept in during the night. Jacob looked up and, for a brief moment, their eyes met. There was a depth of understanding, a silent acknowledgment of the journey each was on. For Sarah, it was a journey of unwavering support and love, and for Jacob, it was the relentless pursuit of a connection to the divine. Thinking of love, he realized how he longed to cleave to God, to be His pampered child; a child, not a servant.

When Jacob entered the kitchen to prepare a cup of coffee, he was surprised to find Sarah still up, diligently mending his socks and underwear.

"You should be in bed," he chastised her caringly.

"I couldn't sleep. When are you going to bed?" she asked.

"Soon," Jacob responded vaguely.

Jacob's eyes shifted to the kitchen clock; it was twenty-five minutes past midnight. This reminded him of a meeting with the Rebbe. He had sought advice on whether to focus on mathematical research or dedicate himself solely to Biblical study. The Rebbe had responded with gentle amusement, "My father-in-law, the Previous Rebbe, once said that the first hour after midnight is a silly hour, but such silliness I never heard! If I would have known why you wanted to see me, I wouldn't have given permission. Now go in good health, carry on with your mathematics, and you will be successful both in Bible study and in your mathematical research!"

Sarah had finally gone to sleep. In the quiet of night, Jacob chanted a Chassidic melody while examining the flow of HD numbers, each the result of billions of calculations, indicating the computational analysis involved in one sequence of Hebrew letters tested against the sequence X. Now and again there was a pause between displayed HD numbers as the processor laboured under the gruelling load of computation.

White icy spider webs lay flattened against the window glass behind the screen. Jacob shivered as a cold draft entered from the loose window frame. Reaching for a blanket from a nearby milk crate, he wrapped it snugly around himself. The grey blanket induced memories of photographs of Klondike prospectors huddled around a camp fire and, for a fleeting moment, Jacob saw himself knee-deep in a river, panning for gold. But instead of gold, he sought to unearth Biblical texts that corresponded to the DNA of exon X. These texts were the shining yellow nuggets he was sifting from the sandy embankments, his goal nothing less than to expose the Bonanza Creek concealed in the book God had bequeathed to mankind.

The room dimmed as a light bulb flickered and died. Jacob squinted at his screen, the text barely visible. It struck him how the loss of a mere light bulb could compromise his epic project. Self-doubt crept in. Was he merely chasing illusions? Did he not know that his project was a gamble and, like the gold prospectors he so admired, he was a speculator, naively believing he could pick up textual gold as soon as he set foot on the Klondike Trail of the soul?

From the floor below, raucous laughter was mixed with a song: "Don't rock the boat!" Jacob wondered if his algorithm was rocking the spiritual boat in Heaven. Although his intentions were noble, was there an underlying arrogance in attempting to extrapolate human reasoning to divine logic? He was struck with a terrifying thought - what if human logic was entirely disconnected from God's logic? But then scientific marvels such as Quantum Theory or the General Theory of Relativity would not exist. There would only be the chaos before creation.

Needing a momentary escape, Jacob stepped onto the porch, taking in the crisp, damp air, hoping it would reinvigorate him. The city lights, distant and subdued, barely illuminated his surroundings. Above, a jet's contrail painted a stark line across the moonlit expanse. He took a deep breath and returned inside. A laundry basket full of dirty clothing lay near the bathroom door. Often in the middle of the night when he could not keep his eyes open, Jacob took the laundry down to the basement where two coin operated washers and one dryer were available for the tenants. He decided to go now. The exercise would keep him awake the rest of the night. He locked the front door and went down stairs.

As he loaded the washing machine, a mirror caught his reflection. The lines under his eyes resembled skate marks on a frozen lake, and three of them formed a distinct right-angle triangle. This brought a smile to his face as it called to mind Shmuki's famous refutation of the Pythagoras Theorem. When Jacob first discovered the theorem in high school, he shared his excitement with Shmuki. A week later, Shmuki claimed he had performed a scientific experiment that disproved the theorem. He had drawn a right-angled triangle in a sandbox with his shoes, then measured 3 shoe lengths in one direction, 4 shoe lengths in the perpendicular direction, and finally precisely measured the length of the hypotenuse, toe to heel. Instead of getting exactly 5 shoe lengths as Pythagoras claimed, Shmuki discovered the length to be 5.2 shoe lengths. Hence the error in the Pythagoras Theorem which had gone unnoticed for three thousand years.

The young Nigerian woman walked into the laundry room, carrying her infant son in one arm and a garbage bag filled with laundry in the other. Her silk, variegated dress shimmered as she loaded the washing machine. Meanwhile Jacob took the initiative to hold the baby, gently running his fingers through the child's tight, curly black hair. The woman smiled affably, possibly recognizing him as the kind stranger who often left bags of food on her door handle. Her washer's spin cycle ended with the triumphant thump of a Cossack dance. She took the baby from Jacob and left the room.

Exhaustion settled over Jacob. He leaned his head against the wall, drifting into a dream in which a vaccine for the corona virus had been developed. The President of the United States had entrusted Jacob with the task of inoculating the children of New York City. On the FDR Drive, Jacob parked his car amidst the traffic, scaled the barrier, and proceeded towards Harlem with a backpack full of vaccines. The streets were quiet, but now and again, a gust of anger blew through the lanes as gun shots rang out in the night. Cupping his hands around his mouth, Jacob called to the children in the apartment buildings: "Vaccine! Vaccine! Please come get the vaccine. It won't hurt and will protect you against the corona virus. I have lollypops: red, green, and blue. Choose whichever you like. May each one of you grow up to be a joy to your parents and your community. I love each and every one of you. May God bless you all."

CHAPTER
FIVE

RETURNING TO HIS APARTMENT, Jacob went to the bathroom. Standing over the toilet bowl, he looked down and from the bifurcating stream of urine he read the story of his life - the struggle between religion and science. Two large bubbles, side by side, the eyes of a flat-faced exotic fish from the deep, floated on the yellow mattress of tiny bubbles. A drain fly clung to the door of the medicine cabinet. Jacob squashed it with a piece of toilet paper. As he was about to drop the crumpled paper into the toilet he hesitated. How could he urinate on God's creation? He turned to toss the paper into the trash bin beneath the sink and wondered if he could still consider himself to be a sentient being. As a result of according a respectable burial to the insect, he soaked his boxer shorts and the cover of an old Scientific American publication that lay on the floor between the toilet and the sink.

Jacob had to change his underwear. He pulled open a drawer in the dresser and changed in the darkness. Sarah was asleep, wrapped in a white sheet like a fillet of sole on a fishmonger's counter. The bedroom was cold. Each room had an electric baseboard with its own thermostat. To economize, Jacob usually turned off the bedroom heater during the day. At night he lowered the temperature on the other thermostats to 50 degrees and raised the one in the bedroom to 68, but tonight he had forgotten to do that. Sarah lay in the cold room, her exposed veined feet illuminated by the closet light she had left on for Jacob.

After adjusting the thermostat in the bedroom, Jacob left the door ajar so he could hear Sarah's breathing but at the same time retain most of the heat in the bedroom. Moments after returning to his desk, sleep enveloped him, his face cushioned by his hands, elbows stabilizing his slumber. A sudden jolt of electricity streaking to his hand from the funny bone in his right elbow, rattled his brief respite. Grimacing, he rubbed his hand until the pain subsided. Once the pain receded, his eyes once again succumbed to weariness, and he was cast into a dream from his adolescent years of a football game at Molson Stadium. He was standing on Pine Avenue with friends, hopeful eyes scanning for someone with an extra ticket to give away. A benevolent middle-aged man emerged from the crowd streaming into the stadium and handed Jacob a ticket. Awakening with a start, Jacob wondered who that man was and why would he choose a scrawny boy with ear locks and a skull cap as the recipient of his generosity? Jacob envisaged that this is how the Messiah will reveal himself, completely unexpected, a beaming smile on his face, holding the ticket to salvation in his hand.

A large integer number attached to the tip of a spear appeared on the screen. Jacob recognized it as a prime number. He thought for a moment, then keyed in a larger prime number. And so the computer hurled numerical spears at Jacob, and he in turn, responded with larger prime numbers until, at last, Jacob proved that one of the computer's numbers was not prime at all by invoking the 'AKS primality test.' Jacob had prevailed. Encouraged by the minuscule victory, he returned to the algorithm with renewed conviction that it would soon guide him to his target.

Moments later, Jacob heard a strange, flapping noise from the kitchen, followed by a flurry of activity, then silence. He stood up, opened the light in the kitchen and saw a black bird flying in circles. He immediately recognized a large bat. His first thought was to close the door to the bedroom where Sarah was sleeping. On the way back from the bedroom, he took hold of a broom that stood against the kitchen wall. Swinging the broom back and forth, he thought he had cornered the bat, but flapping its wide umbrella wings, it suddenly swooped down on him, scratching his neck. Jacob struck a wing with the broom but the bat continued flying. The broom head was not rigid enough to inflict injury. He opened the porch door, then stood in the entrance to the kitchen trying to force the bat toward the porch door from where cold air was filling the apartment. He advanced cautiously, waving the broom. The bat vaulted to the ceiling. For a moment Jacob lost sight of the bat, then found it hanging upside down from the light fixture. With gesturing raised arms he thwarted the bat's attempted flight toward Sarah's door. There was a fissure in the ceiling exposing the rotten wood joists supporting the roof. The bat flew in diminishing circles around the fissure, then fluttered directly into it. Jacob hurried to the living room and tore apart one of the cardboard boxes he had used to explain the packing program to Sarah. He stood up on a chair and attached the cardboard over the opening with green painter's tape. Looking around the kitchen, he was dismayed at the havoc wreaked by the bat. In the bathroom mirror, Jacob saw an ominous scratch on his neck, shaped as a cross. The meaning of the encounter with the bat gnawed at him.

After Jacob straightened the table and chairs in the kitchen, he returned to his desk, determined to make up for the time lost skirmishing with the bat. From where he left off, he followed a string of program executions in a window on the right side of the screen, confident that each step weakened the bonds of this final exile of the Jews and brought humanity a step nearer to the final redemption. Each computation brought him closer to the absolute limit of human history in real time. To reach the limit, Jacob required unwavering dedication and the willingness to risk everything, even his life. King David, with a seemingly impossible task, had not known his stone would find the minuscule vulnerability in Goliath's armour. The probability of that happening was infinitesimally small. Only when he was prepared to perish in his seemingly insurmountable quest did God perform an act transcending nature in the guise of the possible.

Jacob reminisced about the beginning of his personal mission to bring the Messiah. It unfolded on the 28th day of the Hebrew month of

Nissan in 1991. Jacob had traveled to Brooklyn to attend a *farbrengen*, a gathering of Chassidic men in a spirit of love and camaraderie. The large synagogue at 770 Eastern Parkway was bustling with thousands of Chassidim. Following an extensive discourse, the Rebbe began distributing whiskey bottles to the attendees. Chassidic melodies were sung in unison as men filled their small whiskey cups, and extended them to the Rebbe, eagerly awaiting a nod and a wish of *L'Chaim* before uttering a blessing and sipping the whiskey. Patiently, Jacob waited for his moment of acknowledgment. When it came and the Rebbe met his gaze, ecstasy filled him. With a nod and an audibly spoken "*L'Chaim*" and to good health, Jacob felt a surprise stir within him as the Rebbe gave him an extra blessing for a healthy life. Resuming the farbrengen, the Rebbe delivered a discourse about redemption, weaving it into the unique spiritual significance of the current year and month. Near the end, the Rebbe addressed his Chassidim with fatherly intimacy, posing a heartfelt question: "What more can I do to urge the Jewish people to demand the arrival of the Messiah? Our exile persists. All I can do now is to hand this matter to you. You must do everything in your power to bring the Messiah, immediately, here and now. I have done all I can: from now on, you must do whatever you can."

Jacob left the *farbrengen* in a state of bewildered contemplation. For four decades, the Chassidim had leaned on the Rebbe to usher in the Messiah, with many believing he would reveal himself as the Messiah. But now the arduous task was being entrusted to the average Jew. Jacob felt a profound obligation to the Rebbe but lacked the scholarly knowledge of Biblical texts. Despite this, if the Rebbe believed he could contribute, then he must hold some capability. But what could he - a mediocre mathematician - achieve? Regardless, complacency was not an option; he was duty-bound to harness whatever few skills he possessed towards the Messiah's arrival.

Dimmed by weariness, Jacob caught sight of a peculiar pattern within the sequences of HD numbers spewed out by the algorithm. There were obvious oscillations around the HD number $y = .297629175316$. This particular value signalled that the X sequence was notably misaligned with Hebrew text sequences, yet its periodic reappearance was curiously undeniable.

An initial suspicion pranced about in Jacob's mind: perhaps y was linked to the well-known irrational numbers, pi (3.141592...) or e (2.718281...). After some consideration, he could not find any obvious connections between these numbers and y. However, an air of familiarity with y lingered in his thoughts, prompting him to delve into his mathematics tower, an abundant source of theoretical contexts and mathematical concepts.

As he skimmed through a book on Dynamical Systems, the name Fibonacci sprung from the pages. The Fibonacci numbers, known for millennia, consist of an integer sequence 1, 1, 2, 3, 5, 8, 13, 21, 34, 55, 89, 144, etc., wherein each number, after the first two, is the sum of its two predecessors. Moreover, a sequence formed by dividing each Fibonacci number by its precursor, 2/1, 3/2, 5/3, 8/5, 13/8,....... converges to an extraordinary irrational number known as the Golden Ratio, approximately equal to 1.6180339887. Although Fibonacci numbers pervade various mathematical contexts and biological phenomena such as branching in trees and the arrangement of seeds in a sunflower, Jacob found no discernible connection to the HD number .297629175316, which suggested that fewer than 30% of the letters in X matched a 959-length sequence of Hebrew letters in the Bible.

Two additional cups of coffee did little to alleviate Jacob's perplexity; the mystery that had ensnared his mind persisted. After several more attempts, the HD numbers advanced, measuring the proximity to X of additional Hebrew sequences, and the number y vanished from the screen. When resolution seemed unattainable, a sudden intuitive spark propelled Jacob toward exploring the reciprocals of the Fibonacci numbers: 1/1, 1/2, 1/3, 1/5, 1/8, 1/13, 1/21, and so on, a sequence converging to approximately 3.35988566662431, known as the reciprocal-Fibonacci constant.

A faint memory surfaced in Jacob's mind, a reminder that this number bore relevance in the field of artificial intelligence. Although it was greater than 1, and therefore could not be a normalized Hamming Distance, a cascade of disappointment swiftly morphed into ecstasy when Jacob discerned that number as the reciprocal of y. He had discovered a splendid result: the reciprocal of the reciprocal Fibonacci constant attracted Hamming Distances near a special 959-length sequence in the Book of Deuteronomy. Tracing back to where HD had converged to y, he identified the pertinent verse in Deuteronomy: chapter 29, verse 28. *"The hidden matters are for God, our God, but the revealed ones are for us and for our children forever."* A thought seeded itself within Jacob. Was the verse a divine warning against probing into the mysteries of creation? Was exploring God's existence scientifically indeed a sin? But Jacob had treaded too far along the Klondike Trail to reverse his journey. His resolve was unshakable, even if it meant venturing into God's sanctuary for which the death penalty is exacted.

In a realm where science and belief intertwined, Jacob stood convinced that the Bible had communicated with him through the universal language of mathematics. Though the message was veiled in mystery, he experienced an exhilarating certainty that he had witnessed something beyond material existence, something supernatural. The choice he had made many years ago on entering university to study mathematics was vindicated. Subconsciously, he must have understood even then, that between the physical world and the spiritual realm lay the abstract language of mathematics - queen of the sciences - straddling the two as a lustrous revelation of the power of logic that God had implanted in the human brain.

Engrossed in his findings, Jacob shuddered when an unfamiliar voice emanated from the computer, "Jacob, Jacob, remove your shoes; you are standing on sacred ground." A blinding light flashed from the screen, and Jacob instinctively averted his eyes. A plume of smoke went up from the tree in the alleyway visible through the window. Promptly, he discarded his shoes and stood upright, his attention riveted to the unexpected supernatural occurrence.

"I am the Angel of the Hamming Distance," the voice intoned solemnly. "I come with a message: your faith in God is deficient."

Jacob, unshaken, responded, "I seek to transcend faith, to ascend to a level where understanding permeates faith."

The angel retorted, "Jacob, offer your prayers with an innocent heart, not tangled with the cunning of your scientific ponderings. In the realm you seek, understanding is absent. Only through the illogical might of belief can you perceive God. You've misled yourself, esteemed professor. Your purported belief in God, requiring a program for validation, is insincere. Jacob, God has given you insights on how to reinterpret time, how to solve the age of the universe problem, and how to explain dark energy. Is that not enough? Do you think you accomplished that by yourself? You give too much credit to the false prophets of your craft. God can handle paradoxes and contradictions. Man cannot. The particle-wave duality of quantum mechanics is beyond man's comprehension, but poses no problem to God. When you will no longer demand scientific proof, God will show Himself to you."

Suddenly, a spectral arm with a hand shaped like a talon, emerged from the screen. With his right elbow on the desk, Jacob linked hands with the angel. There was a 0 mark on the desk to Jacob's left side, symbolizing perfect sequential alignment, and the number 1 to his right side, indicating pure chaos. He touched his skull cap with his left hand while his right hand gripped the angel's. The vertical line between their interlocked hands served as the gauge for the Hamming Distance. With an outburst of determination, Jacob forced the angel's arm toward 0, but not all the way down. Suddenly the arms jerked upright, then tilted toward the right as the Hamming Distance nudged the number 1. Jacob and the angel wrestled throughout the night, the needle oscillating between the 0 and 1. As dawn's light brushed the window's upper edge, Jacob implored, "Please tell me your name!"

"I cannot," responded the angel.

"Then bless me!" Jacob demanded.

"Release me. Then I will bless you," the angel stated. "But my blessing only assures your success in transforming this lower, material world into a dwelling suitable for Godliness."

Jacob replied, "I do not need your blessing for that. I can accomplish that utilizing my scientific arsenal. I need you to bless my algorithm so I can bring the Messiah!"

"You will not succeed in bringing the Messiah because your algorithm stems from your evil inclination disguised in holy thoughts. You rely too much on your algorithm and treat the Bible merely as another data set," the angel said as its amorphous fingers slid away from Jacob's grasp and vanished inside the screen.

The screen turned black. With hysteria in his eyes, Jacob groped the screen for evidence of the encounter, mumbling through tearful whimpers: "Dear God, I don't want to wrestle with Your angel. I want You, only You! Show me Your silhouette. Let it slide across the screen as when Moses glimpsed Your back from within the cleft of the rock."

CHAPTER
SIX

AMIDST THE CHAOS OF WAR, Sarah's family found sanctuary on a farm, a property of a benevolent Polish family, once employees of her father. One day, while her parents took their toddler Sarah in search of food, German soldiers brutally murdered her three elder siblings. Following the war, Sarah and her parents navigated through various displaced persons camps, eventually finding shelter in a bombed building in the suburbs of Paris, where the city gave way to farmland.

In the bitter winter of 1950, Russian immigrants, including Jacob and his mother, settled in the same dilapidated neighborhood as Sarah and her parents. There was no heating or running water. The children were tasked with retrieving water from a distant well in the country-side. It was there that the paths of Jacob and Sarah became entangled. They patiently waited in line, filling their buckets. When they reached Sarah's building, Jacob offered her one of his buckets. Sarah, while thankful, declined and resolved to fetch two more. Jacob accompanied her and helped her carry them back.

A trip to the Canadian Embassy offered a glimmer of hope to Sarah and her family. Her father rubbed her cheeks, eliciting a vibrant red, aiming to portray the illusion of robust health. The officers remained unconvinced, labelling them as too frail for Canadian residency. Yet fate intervened, and several weeks later, visas materialized. The Hoffman Family soon embarked on a journey to Canada. The tiny quarters deep inside the ship were crammed with broken valises and sacks of curdling cheese hanging from the ceiling, fat snow-white bellies of beheaded ducks. On the second day at sea, Sarah stood on the deck with her father as the salty spray of the ocean bit into their faces. Suddenly, a fierce wind wrenched the hat from her father's head. He reached far over the railing trying to catch it. Horrified, Sarah watched him tottering above the heaving sea, one hand on the wet railing, the other straining for the hat. The ship stopped to lower search boats in a failed attempt at rescue in the tumultuous waves of the North Atlantic Ocean. A few days later, a train crowded with immigrants brought Sarah and her mother to Montreal from Quebec City.

Five years after arriving in Canada, Jacob's mother died of stomach cancer. At her gravesite, the fifteen year old youth vowed to become a scientist and cure the disease. Jacob continued his studies in a yeshiva under the tutelage of kind teachers who took a special interest in the orphan. Jacob found science books which he hid from the rabbis and studied secretly in the bathroom. His journey led him to McGill University for mathematics studies.

One day, soon after he graduated, he met one of his former yeshiva teachers. On a street corner, the teacher, a heavy smoker, rolled off the silver paper from the butt of a cigarette and scribbled a telephone number on the inside. A week later, Jacob was reunited with Sarah whom he vividly remembered from Paris and whom he had seen sporadically on the streets of Montreal. He had not dared approach her earlier due to the modesty that ruled the orthodox community. Match makers had proposed fine young men to Sarah, but she had waited patiently for Jacob. Their relationship blossomed into deep and comfortable conversations. Before accepting a proposal of marriage, Sarah voiced a stipulation: a desire for a large family so that they could name their children after her father, siblings, and other lost members of their families. Understanding the weight of her wish, but afraid to quell her aspirations, Jacob replied, "We will be blessed with as many children as God gives us."

The Rebbe conferred his blessing on their marriage through a letter addressed to Jacob. The early years of their union were marred by the heartache of multiple miscarriages. For Sarah, it felt like each one of them was a poignant addition to the six million lives already lost. One occurred in the seventh month. She could not bring herself to look at the tiny sack of genetic matter, which a nurse had quietly ushered away to a laboratory hidden within the depths of the hospital. An aching void became deeply imbedded in her consciousness.

While Jacob sought solace in his research, Sarah devoted her time to overseeing the construction of a summer cottage in the countryside north of the city, a place meant to one day welcome the children they had yet to bring into the world.

Today marked the anniversary of her father's tragic drowning. Since there were no male relatives in her family, the solemn duty of reciting *Kaddish* for her father had fallen upon her shoulders. Passing through the downtown Chassidic neighborhood, Sarah imagined herself strolling along muddy lanes in the dawn stillness of a pre-war shtetl. She turned into a narrow alley and ascended a creaking wooden staircase to a small landing. There she found a discarded samovar on the floor and well-worn prayer books resting on a table. This was the women's entrance to an ancient synagogue. Concealed behind a heavy blue velvet curtain, Sarah stood alone in the upper balcony. Downstairs, men prepared for the morning service, donning their phylacteries and prayer shawls, then raising their arms imploringly toward the heavens beyond the rafters. A chandelier, its light bulbs reduced to a mere flicker, hung low over the steps leading to the Holy Ark. Near the pulpit, elderly men diligently studied the Talmud, their flowing white beards resembling bakers' aprons on a clothesline. This was the very synagogue that Sarah had visited with her mother on the High Holidays. Drawing nearer to the curtain, she listened for the cue, then began to chant softly in Hebrew, tears streaming down her cheeks: "Exalted and hallowed be His great Name...." Afterward, she whispered, "Dear Father, I am still without children. Please, pray for me."

Sarah wiped her face, put on her sunglasses, and stepped out into the morning sunlight. Her kerchief was loosely knotted around her neck. She cautiously navigated past an oil spill that resembled an animal. As she moved along the bustling summer streets, she was acutely aware of her bare feet in sandals and the airy sleeveless summer dress she wore. She was not as observant as the women around her. On a nearby balcony, a woman sat surrounded by her children. Despite the sweltering heat, the woman wore a wig beneath her cap. At a busy intersection, a young Chassidic boy bent over a newspaper dispenser, pecking at a headline while trying to decipher it. Nearby, an elderly man with a stooped posture, dressed in a black bowler hat and a floor-length black caftan, crossed the street. A scantily clad young woman approached him. For a fleeting moment, from Sarah's perspective, the woman blocked the man from view, and in that ephemeral juxtaposition, Sarah perceived a message emblematic of modern times.

Amid the noisy traffic, a young girl reached out and touched Sarah's elbow. The girl appeared to be around ten or eleven years old, her short black hair neatly tucked behind her ears, her pretty face exuding innocence. She offered a bag containing two candles, a calendar of Sabbath candle-lighting times, and a small Book of Psalms.

"No, thank you," Sarah politely declined.

"Are you afraid to discover that you're Jewish?" the girl asked, her smile possessing a wisdom beyond her years.

"How do you know I'm Jewish?" Sarah questioned, resisting the jostling of the crowd behind her.

"I can see your Jewish soul," the girl replied matter-of-factly.

"What does a Jewish soul look like?" Sarah asked, intrigued. The girl responded with a mysterious smile before saying, "Please take the Psalms. They will help you with whatever you need."

Standing in the bright sunlight atop a wooden exterior staircase, Sarah paused before ringing the doorbell. Peering through a slender gap between the window curtains, she saw an elderly woman, shrouded in a loose floral dress, gradually descend the dim staircase, her thick, arthritic fingers protruding from lengthy sleeves.

"Mother, how are you?" Sarah asked, their cheeks brushing during an embrace. They ascended the steps together, stepping into a soft light at the landing. A photograph of Sarah's father, taken moments before boarding the ship to Canada, hung on the wall. He was in his mid forties then, Sarah's age now. In a mirror nearby, she compared their faces, searching for semblance in the ragged tapestry of her childhood memories.

In the summer, Sarah and Jacob sought refuge at their cottage, away from the tumult of the city. She would pick him up from the university in the late afternoon. Past the bridge out of the city, the traffic dissipated quickly. From a field to their right, a flock of starlings lifted off, a magnificent murmuration, folding, expanding and contracting smoothly in waves of movement across the highway.

The city's tension melted away as they navigated through the mountains, long mud sloughs - surprisingly present in July - staining the vibrantly verdant slopes. Jacob's relaxed expression suddenly changed upon spotting a dead raccoon by the side of road. Sarah observed his countenance grow sombre.

"My precious Jacob," she said gently. "Not every dead raccoon symbolizes a slain child." He was taken aback, surprised that she would say that, but remained silent.

The town of St. Agathe appeared to their left before being swallowed by an overpass. Ahead, the undulating silhouette of Mount Tremblant formed a horizon of overlapping bell curves and peaks. Sarah sought the beacon atop a hill, their cottage nearby. Leaving the highway, they followed a secluded road, passing by markers of local life before the gravel underfoot signalled their arrival amidst the piney depths. During the last stretch, Sarah hoped that one day, in the near future, they would arrive with children. As she and Jacob entered their cottage, sunlight filtered through expansive kitchen windows. After supper, Jacob sat at the kitchen table immersed in his research, while Sarah, having finished folding clothes, sought solace outside. The moonlit mountain slope was framed by the patio's eastern screen, and insects, entranced by the silent blue glow of the electronic bug zapper, clung to the mesh. Gently, Sarah tapped the screen, liberating them from their fatal attraction, then lay down, closing her eyes to the lulling sounds around her. In the early morning gossamer webs of dew lay strewn across the grassy grounds. Sarah was careful not to step on them as she strolled about barefoot, soaking her feet.

Jacob had to attend a lecture at the University of Montreal. They drove through fog-enshrouded mountain peaks. South of St. Jerome, the terrain became flat, good farm land. The speeding cars slowed down as they glided through the ever denser conurbation, an endless military convoy directed at the city center sleeping beneath Mount Royal.

CHAPTER
SEVEN

IN THE EARLY hours of morning, Jacob lowered his head to his folded arms resting on his desk and dreamt he was in an amphitheatre witnessing a conversation between the Rebbe and Einstein.

Rebbe: 'Shalom, Professor Einstein. It's an honour to meet you. Your contributions to physics have truly transformed our understanding of the universe.'

Einstein: 'Thank you, Rabbi. I must say, I'm intrigued by the intersection of science and spirituality. Your work in guiding and inspiring your community is also quite remarkable.'

Rebbe: 'Science and spirituality, while seemingly distinct, share a deep connection. Just as your theories have revealed the underlying order of the cosmos, our Kabbalistic teachings illuminate the spiritual order of the world.'

Einstein: 'While I don't adhere to any particular religious belief, I do recognize the profound impact faith and philosophy have on shaping human values and society.'

Rebbe: 'Indeed, the moral and ethical compass that faith provides, can guide humanity in applying scientific discoveries responsibly. Without a sense of purpose and ethics, knowledge can be misused.'

Einstein: 'You raise a valid point. As science advances, so does our responsibility to ensure its applications are used for the betterment of all, rather than for harm. But, Rabbi, I've often wondered about the nature of time and space. How does your faith address these concepts?'

Rebbe: 'Time and space, as you've explored, are fundamental aspects of our reality. In our teachings, we believe that these dimensions are not only physical, but also spiritual. Time offers us opportunities for growth and transformation, while space is a canvas for the development of our actions and relationships and most importantly, for our service to God.'

Einstein: 'Your perspective adds a new layer to my understanding. It seems you're implying that time and space are not just abstract concepts, but hold deeper significance for the human experience.'

Rebbe: 'Precisely. Just as you've delved into the mysteries of the universe, our tradition encourages us to delve into the mysteries of our spiritual existence, seeking meaning and purpose beyond the superficial.'

Einstein: 'I've often marvelled at the elegance and simplicity underlying the laws of physics. Do you believe there's a similar elegance in the spiritual aspects of life?'

Rebbe: 'Absolutely. Just as elegance and simplicity often point to deeper truths in science, our spiritual teachings emphasize the essence, the core values that guide us. Finding those fundamental principles and living by them can bring about profound transformations in individuals and communities.'

Einstein: 'Your insight resonates with me. It seems we are both driven by a pursuit of understanding the fundamental truths that govern our respective domains.'

Rebbe: 'Indeed, Professor Einstein. While our paths may be different, our shared quest for knowledge and understanding can bring about a harmonious synthesis between science and spirituality, enriching the human experience as a whole.'

Einstein: 'I believe you're right, Rabbi. Our discussions today have given me much to contemplate. Thank you for sharing your wisdom with me.'

Rebbe: 'The pleasure is mine, Professor Einstein. Let us continue to explore the realms of knowledge and inspiration, each contributing in our own way to the enhancement of humanity.'

. . .

Early Thursday morning, Jacob stood at the porch door, observing the first rays of daylight creep upward from behind the Jewish school. Above the lane, squirrels darted back and forth across ice-coated power lines. He peeked into the bedroom and saw that Sarah was sleeping. Too tired to trek the kilometre and a half to the synagogue for morning services, he decided to stay home. Besides, this would give him an extra 2 hours of work. Jacob twirled the large prayer shawl around his head while reciting the appropriate blessing. As he put on his phylacteries, first on his left arm then on his forehead, he remembered the words of the *Zohar*: *The time of prayer is a time of war with the evil inclination.* He was careful to cover the freckle on his left forearm with the strap of his arm phylactery, as if he could conceal his visible as well as his invisible shortcomings from God. After collecting the four corner tassels of the prayer shawl into his left hand, Jacob recited the *Shema*. His thoughts hovered above the prayer book, scarcely engaged with the meaning of the words he was pronouncing perfunctorily.

There had been occasions in the past when Jacob had prayed earnestly, his sincere supplications propelling him away from earthly existence. Then he had broken through earth's spiritual gravity, soaring into mystical realms, flung deep into the cosmos by the sling-shot effect of the heavy Kabbalistic spheres. Drifting deep in space, he saw a shadow of the Almighty drip-feeding divine wisdom, letter by letter, into earthly existence: first the letter *aleph*, then the letter *beth*, and so on, then arranging the 22 Hebrew letters in the Biblical text to satisfy the myriad constraints that permitted creation and the existence of the universe.

As Jacob meditated on the greatness of God, it occurred to him that existence is the solution to an infinite dimensional optimization problem. This thought agreed with a planned project to present the Bible in a scientific format. It would start with a sequence of axioms:

Axiom 1: In the beginning God created the heavens and the earth.

Axiom 2: Life is the potential to reveal Godliness in the world.

Axiom 3: Repentance is a universal law of nature.

Jacob grappled with The Second Law of Thermodynamics, which

asserts that natural processes are irreversible. If a vase crashes to the floor, the process cannot be reversed. The shards of the vase do not fly upward to the table, re-assembling on the way. But Jacob knew of one counter example: repentance reverses events by transforming transgressions to merits. Hence, Jacob tentatively formulated

Theorem 1: Repentance refutes the Second Law of Thermodynamics.

Before Jacob undertook the Bible-DNA project, he considered trying to write a commentary with the latest scientific findings to go along with the text of the Bible. The first verse of Genesis would read:

In the beginning *there was only the primordial light, particles without mass travelling at the speed of light, c. God breathed the Higgs Field into the empty space outside of Himself. Matter came into existence, possessing velocities less than c, and time began to flow. Then* **God created the heavens and the earth.**

The morning prayers usually took Jacob an hour, but today he completed them in 20 minutes. There were rabbis who had been exempted from prayer altogether due to their assiduous Bible study. But really, Jacob thought, could he compare his labour on the algorithm to Bible study? How could he even entertain such a comparison, knowing full well that science is no more than the chaff of biblical deliberations!

At the breakfast table, Sarah sensed Jacob's anxiety as he could not eat. She said sadly, "How I wish you had never gotten involved with this project." She touched her nose with the edge of her apron. "The plague God has inflicted on the world, when will it end? People are dying and so many are very sick." Jacob nodded in accord with her sentiments, but in his heart he felt confident that soon his algorithm would bring about a vaccine that would end the pandemic and solve the world's many other problems.

Sarah wiped the tears from her eyes with the sides of her hands. Even now in old age she felt insecure for having had only one child, while the women of her age group had many children. Whenever Jacob sensed her uneasiness, he reminded her: "Our Matriarch Sarah had only one child, Isaac, and look what a great nation came from him." He embraced her and assured her that he loved her. To further placate her, Jacob swallowed a tablespoon of porridge and chewed on a piece of red pepper. A hair from his untrimmed moustache was caught between his upper front teeth. He felt it with his tongue and against the inside of his upper lip and it annoyed him. He thought, 'Here I am climbing to the greatest heights of spirituality, yet my ascent is hampered by a hair.' He modeled this conflict between high aspiration and lowly distraction as a product XY, where X, the goal of his algorithm, approached infinity while Y, the hair, approached minus infinity. The product of these divergent forces, XY could be any number. Added to Y was his molar pain. When he had retired from the university, his dental insurance plan was cancelled. Sarah suggested they apply for an interest free loan to pay for the dental work Jacob urgently needed; but he could not see himself walking around the chapel, as he had seen others do, holding the application form, begging for the signatures of three guarantors. He would rather wait until the pain was unbearable, then have the tooth extracted.

Mid morning Jacob sat staring at the numbers flitting down the screen. Sarah approached him with the re-warmed porridge he had left over at breakfast and spoon fed him like she would a toddler in a high chair. Jacob had no idea what he was eating nor whether it was hot or cold. He was deep in thought about the standard model of modern physics which claimed that there are 12 types of fundamental particles called fermions: 6 quarks and 6 leptons. These particles live in a space-time of 10 dimensions: 1 dimension of time and 3 dimensions of space, where at each point of 3 dimensional space there is an additional tiny 6 dimensional space structure called the Calabi-Yau manifold. Thus, there are 12 fundamental particles functioning in 10 dimensions, defining in total 22 fundamental objects. These 22 objects corresponded to the 22 Hebrew letters, the basic building blocks of creation. With this simple dimension analysis, Jacob had further evidence that his project was on solid footing.

Unwittingly, he had consumed the entire bowl of porridge. Sarah was happy. She went to the kitchen smiling with accomplishment as Jacob, oblivious to everything around him, continued his work. He recalled the hype associated with Bible codes in the seventies and eighties which sought geometric patterns in the Bible. For Jacob, those searches had lacked a scientific basis. What he had gleaned from those futile efforts was the simple perception that codes were hidden in the Bible. Years later, when he saw one of his grandchildren studying biology, it occurred to him that there might be a connection between Bible codes and DNA codes. Every scientific quest starts with a hunch, but Jacob had more than a hunch; he had enough circumstantial evidence that the codes of life were first written in the Bible, then transcribed to the genome.

Early in his research, there was an encouraging discovery that Jacob made. The 26,564 genes in the human genome contain 233,785 exons. When testing his packing program with DNA sequences that are not exons, the algorithm failed to find partitions of the 22 Hebrew letters into 4 sets that had the same probabilities of occurrences as A,C,G,T in the DNA sequence. The only conclusion that could be drawn was that there are no matches between non-exon DNA sequences and Hebrew texts. It made perfect sense: only exons possess the potential for matches with texts from the Bible. This meant that God did not waste precious text in the Bible on sequences that were not needed for protein production.

During the winter of 2020, when news of the pandemic began emerging, there were days of despair for Jacob, fearing there was a fundamental flaw in the design of his algorithm. The program was close to the end of the fourth book of the Bible without finding a single Hebrew sequence that perfectly matched the DNA sequence of exon X. This did not shake his confidence that the sequence existed, but rather raised concern that his packing program did not reveal all the perfect packings of Hebrew sequences. If some had been missed, then the exact match of a Hebrew sequence might be among those and would be tested by the comparison program. There were other explanations for failure. One was that instead of using consecutive Hebrew letters, God might have used every second or every third Hebrew letter in creating DNA sequences, as in the story of Avner, a 13th-century

Spanish Jew who had renounced his faith. Avner's mentor had been the Ramban, a prominent medieval scholar. During Yom Kippur, the most sacred day of the year, Avner summoned guards to escort the Ramban to his palace. There, he brazenly feasted on pork in front of his former teacher.

"How did you descend to such depths?" the Ramban lamented.

"You bear the full responsibility! You once taught the Bible section Haazinu, asserting that within its 52 verses lay the names of all Jews who ever existed. Despite countless searches, I never found my name within the text," Avner retorted.

"My claim still stands," the Ramban replied firmly.

"Then show me where my name is!" Avner challenged.

In response, the Ramban implored the heavens for clarity. Then, turning to Avner, he directed him to verse 26, which reads: "I (God) declared in my heart that I would disperse them, erasing their memory from humanity."

"What are the third letters of each word?" the Ramban asked.

Avner's face drained of color. The third letters of four words from the verse collectively spelled his name.

Such a textual coding for creation, Jacob thought, would be unnecessary as there are sufficient sequences of consecutive Hebrew letters to include all the exons. Any other pattern for Hebrew sequences would be an unnecessary and devious scheme. Jacob was convinced that it would not cost God anything extra to have created life in the straightforward way Jacob premised in his algorithm.

With a cup of coffee in hand, Jacob resumed checking the packing code for errors that might have caused it to miss partitions of the Hebrew alphabet. As he worked he spoke to God, "If indeed You are omnipotent, You can contract Yourself to a two dimensional image on my screen. To be infinite only is a deficiency; to be truly infinite must mean to be finite also. I ask merely for a 2 dimensional peek on my screen. Those who understand Your Holographic Principle know that the 3 dimensional world You created is defined by the 2 dimensional book, the flat pages of the Holy Bible You wrote. It is this 2 dimensional world that is prominent, not the illusion of three dimensions."

Jacob could not deny the physical three dimensional world of DNA life and its embedding in the two dimensional algorithm simulated on his computer screen. The two worlds - reality and its simulation - were entangled, connected by a wormhole. Jacob was searching for the wormhole to heaven.

There were times when Jacob wondered if he was delusional to think he could actually bring the Messiah with his algorithm. Whenever he felt discouraged, he brought to mind a letter he had written to the Rebbe when he was a young professor, requesting a blessing for success in his research. The Rebbe wrote back: "May your efforts take you beyond the natural and bring the greatest blessing of all: the coming of the Messiah, may he come soon." These words encouraged Jacob to strive for important scientific results. And so he placed his hand on his forehead and wiped away all Heavenly decrees against his algorithm.

Over the years Jacob had not spared any effort to fulfill the Rebbe's words. His first attempt to use mathematics beyond traditional areas of research was related to the concept of time. He disliked the notion of continuous time: such a concept attributed to humanity more control of life than Jacob thought possible. The smooth continuity of time implied that man could predict the immediate future just as smooth graphs are extrapolated accurately over short distances. Continuity forbade abrupt change which can occur only in a discrete time, and therefore undermined the power of God. The incident that finally convinced Jacob of time's discreteness occurred while coming back from New York. Sarah was driving and took the wrong fork on the highway near Albany and they were unwittingly on the way to Buffalo. After an hour of driving west rather than north, Jacob realized that the landscape was unfamiliar. A sign on the shoulder of the road finally informed him of their error. Sarah took the first exit and followed country roads back to the main highway north. The narrow gravel roads skirted winding wooden fences and undulating rich farmland. Drooping trees and luxuriant gardens lay behind green painted fences. Cows pastured on a grassy field on the side of the road.

The sight of the posts and the spaces between them intrigued Jacob. At high speed, there was the illusion that the fleeting fence posts formed a continuum. At slow speed, the posts separated out from the continuum and displayed the true discreteness of the posts. As Jacob considered the two images of the fence - one at high speed, the other at low speed - responding to an inspiration, he applied the spatial images to time itself, postulating that each time post had the duration of a Planck time, the smallest unit of time, 10^{-34} seconds, a quantum of time. These quantum times were the supporting columns of time. The time between the quantum times was analogous to the space between the wood posts. From this, he recognized that there were in fact 2 types of time: the time duration corresponding to the widths of the time posts, which he called existential time, and the spaces between the time posts which he called non-existential time. Guided by this idea, Jacob wrote a discrete time version of the Planck-Einstein Equation from which he calculated the existential time of the universe. It accorded accurately with the number given by the Bible, 5780 years. The total time was 13.8 billion years. There was no contradiction between the two ages.

If the physical world was governed by rules of continuity, scientific accomplishments would merely be a sequence of ever improving approximations to truths that were unattainable limits of these sequences. Jacob saw his role clearly, thinking, 'I, the old mathematician, stand alone between the sequence and its limit, straddling the two, brandishing the blazing sword of intellect that God has instilled within me.'

Jacob was pleased that his algorithm functioned in the discrete realm where all results were exact. His algorithm would brook no approximation: the mapping he sought, going from Hebrew text to genetic code, had to be perfect. Jacob approved of God's choice to make it so. The algorithm was a zipper; one side consisted of the X sequence, the other side a Biblical sequence. When no tiny hernias would appear as the zipper was closed up, the HD between sequences would equal 0. To reach this pinnacle of perfection, Jacob had to squeeze the spirituality out of every Hebrew letter into exon X.

CHAPTER
EIGHT

SHORTLY BEFORE 2 PM, Sarah approached Jacob holding a cup of tea in one hand and a plate in the other. Upon the plate lay tiny cookies she had discovered in the pantry, knowing full well that Jacob harboured a fondness for them when they were a luxury they could afford. He would tenderly submerge the hard cookies into the tea, softening them enough to be more palatable.

As Jacob welcomed Sarah with a warm smile, he instinctively began counting the cookies first by twos, then by threes, totalling twelve. In the parched desert, God had counted the Jews multiple times, an expression of divine love. Likewise, Jacob's algorithm, engendering a cataract of near-endless computations, was his expression of love toward God. His brain thrived on numbers. A numerical maestro, Jacob could recite the license plate numbers of the cars parked along both sides of his street. When taking a vision test, it was challenging for him not to remember the numbers from the sequences presented during his prior assessments. Numbers formed an intricate bridge between their need in daily realities and his love for God as expressed in the algorithm.

But why had not the perfect number 0 appeared in the variable HD? Jacob revisited the idea of implementing spaces between words in his algorithm. Initially, during the preliminary studies, he deemed them redundant due to the availability of numerous combinations formed by the 22 Hebrew letters, which could generate all necessary exons for life. Given that DNA sequences lack gaps between bases, he postulated that the Hebrew text should similarly contain no spaces. The spaces were a facilitative tool for human comprehension of the Hebrew text, not a requirement for creation. This logic had steered Jacob toward a decision two years prior. Yet doubt lingered, festering the thought that his assumption could be errant. Moreover, his program could be flawed in other ways; it neglected the five unique Hebrew letters appearing at the ends of words, which he had deliberately omitted in his program's initial version.

Facing his apartment's eastern wall, Jacob recited the afternoon prayers from memory. When Sarah called him for their second porridge meal of the day, there was a knock on the front door.

With shuffled steps, Sarah approached, inquiring, "Who is it?"

"Me, Elka," came a muffled reply.

An embrace was shared in the vestibule between mother and daughter. "Who is watching the children?" Sarah questioned.

"Have you forgotten? Hannah is almost eighteen," Elka responded, raising her hand above her head to indicate her daughter's height.

Elka brought with her a shopping bag filled with a box of blue masks, vials of vitamin D and magnesium, two containers of cottage cheese, and a bottle of milk. Disapprovingly, Sarah shook her head, "There's no need to exert yourself in this weather."

"It's not so bad outside, and the walk is good for me," Elka replied warmly.

"How are you, Tatty?" Elka queried, stepping towards her father to plant a kiss on his cheek, her face still bearing the chill from outside.

"I'm fine, thank God," he affirmed, gazing affectionately at his daughter.

Jacob retreated to his desk. A few minutes later, through the kitchen's back wall mirror, he caught the reflection of mother and daughter engaged in conversation at the table. The sight reminded him of their visit to the Rebbe in the 20th year of their marriage, on a Sunday when the Rebbe, outside 770 Eastern Parkway, distributed dollars and blessings for numerous hours. After a two-hour wait, they stood before the Rebbe. Jacob, head bowed, stammered: "We have no children."

Motioning for him to lift his gaze, the Rebbe reassured, "I give you a blessing for a healthy child."

"Rebbe, please, we need a promise. We are old."

"You will have a child this year," the Rebbe declared in lucid English.

Sarah gave birth at the age of forty-four. Jacob remembered the moment Sarah delivered Elka into the world, akin to how God must have propelled the Higgs Field into the abyss beyond Himself. Sarah, tears streaming with joy, exclaimed: "Elka helped me push! She wanted to come into the world." Jacob mused, 'The universe must have similarly been eager to exist and validate itself before God.'

Jacob recalled that at 12 months, Elka crawled to the large living room window of their home, gazing down at the street below. Jacob wondered what she was thinking as she observed cars and people. Billions of new neural connections must have formed in those moments, as sensory data cascaded into her tiny cerebral cortex through vivid blue eyes. When Elka was 3 years old, Jacob lulled her to sleep with stories. He lay on the recliner, Elka on his chest, her thumb securely in her mouth. Jacob told her the story of a little goat that became lost in the wilderness and could not find its way back to Jerusalem. Elka cried for the goat until her eyes gradually closed. How delightful it felt when, at last, her head grew heavy on him. On her 5th birthday, Jacob taught Elka the geometry of the Hebrew letter aleph: the dot on top is God, the dot at the bottom are the Jews, and the line connecting them is the holy Bible. As she grew older, he taught her the geometric meaning of all the Hebrew letters.

During his graduate studies, Jacob was fascinated by a beautiful theorem that had immediate application for him beyond mathematics. It was a theorem proved by two Russian Jews in the thirties, the Krein-Millman Theorem. He remembered the wording: *Let A be a compact, convex subset of an abstract space. Then any point in A is a combination of its extreme points.* For Jacob, the abstract space was life itself, the set A consisting of a person's experiences; the extreme points of A, the extreme experiences in life. As he sat thinking about the extreme points of Elka's life, he saw her personality shaped by a combination of those experiences.

Glancing at Sarah's subdued expression, Jacob understood that to this day her sadness that Elka did not have siblings had never abated. There had been no older sisters or brothers to help her shed the cocoon of infancy. Elka had been raised alone by a couple almost too old to be parents.

Elka respected her father's research and was hesitant to disturb him, but today she needed to talk to him. A friend of Elka's had been diagnosed with MS and Elka wondered if her father's DNA research could help the young woman. Elka had always relied on her father to devise ingenious scientific solutions to life's problems.

"When the principle of the algorithm will have been established, I'll be able to use very small HD numbers to locate mutations in the exon responsible for MS by comparing the mutated DNA with the associated Hebrew text which is perfect," he explained. "This would guide scientists to change those mutations back to the DNA code at their creation."

"I wish I understood what you're doing," she sighed.

"You do understand what I'm trying to do."

"I do?" she asked dubiously.

"Don't you remember what you wrote in your diary when you were nine years old?"

"No. I don't."

"You wrote: 'Shayna started the fight, but I went to her and told her I was sorry because I want the Messiah to come.' The algorithm is my way of bringing the Messiah."

"Is that really possible?" she asked.

"We'll know very soon."

"I wish you luck."

"I need it."

"I have to go home and prepare supper."

"It's slippery outside," Jacob cautioned. "I'll walk with you."

"No need to, Tatty. I'm careful."

"I've been sitting all day. I need to move," he insisted and went to put on his boots and coat.

They walked side by side in the darkness of deserted streets. Then he thought about his earlier, fruitless attempts to lure Elka into studying science. When Elka was in the last year of high school, she and Jacob walked to a park, on the way discussing infinity. The previous night they had argued late into the night about whether a point takes up space or not. In the playground, Jacob flattened a four foot square area of sand with his shoes as Shmuki would have done, then picked up a twig. With its sturdy tip he drew two concentric circles, one much larger than the other.

"Since the larger circle is longer than the smaller one, there must be more points on the larger circle than on the smaller one."

"Obviously," Elka replied.

Jacob's grin made Elka sense her father had something up his sleeve. Holding the twig like an extended pencil, he drew a straight line from the common center of the two circles, crossing the smaller circle then touching the outer circle. He labelled the point where the line crossed the smaller circle letter A and the point where the line touched the larger circle B.

"Listen carefully, Elka."

"I'm listening, Tatty."

"For each point A on the smaller circle, there is one point B on the larger circle. You see how they're connected by the straight line here. And the other way around as well. For each point B on the larger circle there is exactly one point A on the smaller circle, also connected by the straight line. Do you agree?"

"I agree."

"That means there are exactly the same number of points on both circles!" Jacob said to his puzzled daughter. "You're right to be surprised because didn't we agree that the larger circle has more points than the smaller one?" Jacob's smile deepened the mystery of the apparent paradox. Elka's mouth was agape with curiosity. At length, he said: "My dear daughter, you should know, when it comes to infinity, intuition does not work."

"Wow!" she gasped.

• • •

When Elka graduated from high school, a decision had to be made: attending a teacher's seminary like the majority of her class or going to university to pursue a bachelor's degree in science. In the chassidic community, a year of study in a seminary was a prerequisite for good marriage proposals. Even a short stint in university would discourage potential suitors. Many in the community were of the opinion that the university experience would entice one away from their religious life-style. Ultimately it was agreed that Elka would go to seminary for the first year, then go to university. Her teachers' diploma would facilitate marriage and the university degree would allow her to find an interesting job and help support a future family. Jacob envisaged her life as a superposition of motherhood and scientist. It would be an example of the duality principle he strongly believed was at the very heart of existence itself: living with purpose in the material world, while eyes and heart are directed to heaven.

When Jacob returned to his apartment twenty minutes later, he found Uyai, the Nigerian woman, seated at the kitchen table, her baby sitting restlessly on her lap. Sarah had placed cheerios on a plate in front of the child. Uyai's shyness hindered her from speaking directly to Jacob. Sarah relayed the conversation they had only moments before.

"Uyai's factory had to shut down last week due to a corona virus outbreak. Though she has filed for unemployment insurance, she won't receive her first check for about three weeks. She was wondering if we might lend her a hundred dollars and wrote a note pledging to repay us within 30 days."

"Your letter is not necessary. We're happy to lend you whatever we have," he said with sympathy, then went to his desk where the top drawer contained loose cash totalling 45 dollars. Bringing it to the kitchen, he laid the money on the table.

Sarah remarked, "This isn't a hundred dollars," and took the charity box from the counter. She emptied all its coins into a plastic bag and returned to the table.

Uyai confessed, "I feel terrible. It's for the baby. I don't need anything for myself."

"We insist, Uyai," Sarah affirmed. "It's an honour for us to help. I'm not sure if this amounts to a hundred dollars, but I sincerely hope so. I wish we could offer you more."

Uyai nodded and responded, "I will repay you as soon as my unemployment check arrives."

Jacob gently reassured her, "There's absolutely no need to rush. We're expecting our pension check this week."

Sarah packed a shopping bag with cans of chickpeas and tuna, a box of pasta, and a bottle of milk. She handed the bag to Jacob, who escorted Uyai to her apartment.

Upon his return, Jacob was met with a pained expression marking Sarah's face, a telltale sign of her migraines. He went to get Tylenol from the kitchen cabinet, only to discover an empty bottle. He eyed the clock; it was 8:30 PM, still an hour until curfew. Sarah lay in bed, a cold damp compress on her forehead. In the kitchen, Jacob scribbled a quick note, indicating he had gone to buy paper for the printer and would return home in 15 minutes. He hesitated at the door, realizing he hadn't signed it with his customary, 'Love, Jacob.' Knowing that Sarah would object to him going to buy something for her, he exited the apartment stealthily. Once outside, he walked briskly to the pharmacy.

Jacob entered the pharmacy 5 minutes before the 9 PM closing time. His mask had slipped off the bridge of his nose. The guard standing at the entrance gestured to him to lift it up and to wash his hands with the sanitizer. Jacob did so, then proceeded to the aisle for painkillers. He found the extra strong Tylenol and hurried to pay at the front counter. The short blonde woman standing behind the register pointed to the automated machines that worked with credit or debit cards. Jacob told her that he only had cash. Annoyed by the need to serve him so late in her shift, she took the 20 dollar bill. She wanted to give Jacob the change, but the register had locked automatically at 9 o'clock. Grumbling in a foreign language, the cashier called the store manager to unlock the register. Something about the cashier reminded Jacob of a woman from his past. It was not a pleasant memory that was eluding him. By the time the manager arrived, it was 9:24. A minute after the 9:30 curfew, Jacob left the pharmacy.

Except for a police car in the far corner, the parking lot was empty and the shopping promenade entirely deserted. The lights of the police car in the far end of the parking lot were off to entrap drivers who might be tempted to make an illegal left turn out of the parking lot. As Jacob made his way home, the lone policeman got out of the car and stood directly in Jacob's way. A red tattoo was visible on his neck and circular black ones on the back of his hands. The policeman called out loudly to Jacob: "Stop!" Jacob pretended not to hear. The infuriated officer ran after the old man, grabbed his arms from behind, pulled them behind Jacob's back and steered him roughly to the car while shouting in English: "You bloody Jews brought the pandemic to Quebec from Brooklyn. You don't speak French and you don't respect Quebec laws! The curfew means nothing to you!" As he thrust Jacob against the car hood, blood gushed from his nose through the mask. Holding Jacob bent over the hood, the officer cursed in French. As Shmuki came to mind, Jacob raised his right leg and thrust his shoe backwards into the policeman's groin. The officer released his hold, collapsing into the snow. Squeezing the blood soaked mask tightly against his nose, Jacob walked away with a steady stride. He waited for the street light to turn green at the corner, then crossed Van Horne Avenue and proceeded into the dimly lit footpath crossing the park.

CHAPTER
NINE

SARAH WAS asleep when Jacob returned to the apartment. After discarding the note he had left on the kitchen table, he washed his face in the bathroom, then packed cotton balls into his nostrils to stem the bleeding. He poured a glass of water for Sarah and left it at her bedside with two Tylenol tablets.

Tonight he did not have to worry about falling asleep; the pain in his nose would keep him awake. While the packing program ran in a window on the screen, he checked the news in another window: the corona virus was ravaging old age homes in New York, schools were closed, and employees were told to work from home. Countries around the globe had imposed border closures.

A few minutes later, an errant email appeared from a German Alzheimer's Society. Jacob tried to unsubscribe, but it was encrypted in such a way that made it impossible to delete his email address from the subscription list. As he struggled to eliminate the obstinate email, he unwittingly read that the Society had launched a new GoFundMe campaign for a research project to preserve the memories of SS veterans.

. . .

Jacob felt a frigid wind seeping into the apartment. He thought that it would be on a night like this that he would die. One by one, sensation in his limbs would cease, legs first, then arms. Then his soul would separate from his body with much pain, a fine tooth comb yanked out of a cotton ball, finally severing the bond between flesh and soul.

He stepped to the porch door and looked outside through the window. A strong wind was sandblasting the back wall of his building, bending the tree crown wildly. Jacob cherished inclement weather; through the lightning, thunder, and wailing wind, he heard the slow, steady footsteps of the Messiah.

Jacob picked up the manila folder on his desk and turned to a dog-eared page to read the notes he had prepared for the article he hoped to submit for publication. He skimmed down a page written in a small script that only he could read, "Since exon X is too long for the purposes of exposition, I will employ a short exon consisting of 156 bases, possessing 49 A bases, 42 C bases, 47 G bases and 18 T bases. For comparison, we use the first 156 letters of the Bible. In this text, the Hebrew letter *aleph* occurs 20 times, *beth* 7 times, *gimmel* 0 times, *dalet* 1 time, *hey* 20 times, *vav* 20 times, *zain* 0 times, *chet* 4 times, *tet* 1 time, *yud* 22 times, *kaf* 3 times, *lamed* 9 times, *mem* 12 times, *nun* 4 times, *samech* 0 times, *ayin* 2 times, *pey* 3 times, *tzadi* 2 times, *kuf* 1 time, *reish* 13 times, *shin* 4 times, and *tav* 8 times. There are 22 small packages which we combine into 4 groups, each having the same number of occurrences as DNA bases A, C, G, T have in the 156-length exon. Working with paper and pencil, I found a perfect partition of the 156 length Hebrew text into 4 sets: A' consisting of the letters *tet, kaf, nun, beth, mem,* and *yud* and the total number of occurrences of these letters in the 156 length Hebrew text is 49, which equals the number of occurrences of base A in exon X; C' consisting of *tzadi, hey, aleph,* adding up to 42 occurrences in the Hebrew text which equals the number of occurrences of base C in X; G' consisting of *kuf, daled, shin, tet, tav, lamed, vav* with the number of occurrences of these letters in the Hebrew text equal to 46, the number of occurrences of base G in X. Finally, T' consisting of *ayin, pey, reish,* and the number of occurrences of these letters in the Hebrew text is 18, the number of occurrences of base T in X. There are many other possible combinations of Hebrew letter groupings having exactly the same occurrences as A, C, G, T in the exon. The job of the packing algorithm is to find them all. Now that I have manually found a 4 letter Hebrew alphabet consisting of groups A', C', G', T', the next step is to compare the 156 length exon sequence of A, C, G, T DNA letters with the first 156 letters of the Bible. The first letter of the Bible, *beth*, is in A', the next letter, *reish*, is in T'. Moving along the exon sequence of DNA letters from position 1 to 156, the program moves simultaneously along the corresponding Hebrew sequence: whenever an A in the exon does not overlap with a letter from the group A' of Hebrew letters, the number 1 is added to HD. When there is a match, there is no change in HD. After the Hebrew text and X are compared for all 156 positions, the final HD is divided by 156, yielding a number between 0 and 1, the normalized HD, which is a measure of the distance between the fixed 156 length sequence of DNA letters and the first 156 Hebrew letters of the Bible. HD = 0 means a perfect match, that is, every occurrence of A in the exon matches exactly with a Hebrew letter in A', and similarly for C', G', and T'. HD = 1 means a complete mismatch; that is, at every position of the DNA sequence, the base does not match with the corresponding Hebrew letters. After all the HDs are calculated for all possible perfect partitions of the 22 letters for a given sequence of Hebrew text, the smallest normalized Hamming Distance is displayed on the screen. The Hebrew text is then shifted by one letter to the left, now starting with *reish*, the second letter in the Bible, and continuing for the next 156 Hebrew letters. Once again, the perfect packings are found, then the minimal HD computed. The process of shifting to the left continues more than 300,000 times until all consecutive Hebrew sequences of length 156 are tested against the exon."

Toward dawn, Jacob reviewed the *greedy approach*, a graph theoretic method that produced thousands of solutions, but he did not know if it produced all solutions. For a long exon, the corresponding Hebrew sequences could be packed in a multitude of different packages A', C', G', and T' which possess the same probabilities of occurrences as A, C,

G, T in exon X. It was impossible to test all the packing possibilities manually. Jacob could not determine if there were solutions missed by the packing program. Such a missed packing could be just the one that results in an exact match with the X sequence.

While Jacob was in the kitchen preparing a cup of coffee, he heard a sound from the computer speakers. It was an announcement he had written into the algorithm: "Getting closer." He poured the steaming coffee into the sink and rushed to his desk. Standing bent over the keyboard, he shook the mouse to direct the cursor to the top of the left window and quickly discerned that the HD numbers had made a concerted move toward 0, triggering the voice message. But after a few more calculations, the Hamming Distance veered away from 0. It was a false alarm. With a grimace of distress, Jacob clenched his beard in his right hand and held it tightly when he suddenly remembered that he had not programmed the word "getting" into the code. It was not his style to insert an unnecessary word. The word "closer" was sufficient to alert him. On scrolling down the source code in the right window, he saw that both words were indeed there. Had his memory failed him or had the AI module been mysteriously rewritten?

Jacob recalled a story he had read about an arrogant woman who had been a mediocre artist. She spent her life lusting for fame but no one appreciated her work, so she invested all her money and energy into her website. After she passed away, the AI software an engineer had designed for her, allowed her to live on posthumously. Her website supplanted her life. Jacob thought that perhaps the AI he had coded into his algorithm would allow his algorithm to live on without him, discovering new research problems to investigate and execute. With varied digital incarnations, his algorithm would live forever. But now Jacob was alive, and his job was to make the algorithm work in real time and real space. Committed to the success of the algorithm, he entered the gaping jaw of the source code in hot pursuit of an error, a cavity in a tooth of the Biblical Leviathan. After a thorough examination of the code, he was saddened that he had not found a single bug. Had he located an error, it would have given him renewed hope that the corrected calculations would lead to success, although it would take another year to recalculate all the Hamming Distances. The program was at the 300,119th Hebrew letter of the Bible. There were only a few thousand left shifts remaining. The thought of the program's completion without success steeped the old professor in dread.

CHAPTER
TEN

WHEN JACOB STEPPED into the kitchen Friday morning, he found Sarah vacuuming the floor. He was sad to see Sarah bent over struggling with the old, noisy machine. There had been a time when they could afford a cleaning lady, but Sarah would not hear of it. The idea of someone doing her housework was abhorrent to her. Jacob knew Sarah would become the worker and the cleaning lady the supervisor. It was the same with charity, done with such servile humility that Sarah, as donor, could not be distinguished from the recipient of her donation.

When she noticed Jacob's bruised nose with bloodied cotton protruding from his nostrils, she gasped. He forestalled her shock by telling her that he had gone out for a walk in the evening and had slipped on the ice.

"I left a note on the kitchen table," he said, as if that oblique comment explained his nose and bruised face. "I'm perfectly fine."

"You have to see a doctor right away."

"Where would one find a doctor in these times?"

He offered to help with the cleaning, but Sarah adamantly refused. She would not take a second away from his research time. "How are you feeling?" he asked.

"Thank God, better. I took the Tylenol. Thank you for bringing it."

Tears welled up in her eyes.

"Why are you crying?" he asked, touching her forearm.

"I'm so worried about you."

She knew he was hopelessly addicted to his program. Her only consolation was that his ardent commitment kept him active and prevented him from falling into depression.

"I'm doing what I've been doing all my life - working hard."

"You're older now and there's something dangerously different about this project."

"I can step away from it if I had to."

"I know you better than that. You're too passionate about this project. But I don't want to lose you. I love you. You're all I have. Elka has her own life."

He took her into his arms. She was as light as a child.

"You won't lose me," he assured her. "Only a few days left and the algorithm will come to an end one way or another."

Jacob nodded reassuringly and thought that at their age, they communicated better with silence than with words. They held each other in a long, warm embrace. At that moment he recalled that when he was struggling to define an interesting mathematical problem or to prove a theorem, he would say to Sarah, 'Please tell me what's troubling me,' as if she could penetrate his brain and there, among the frothing chaos, pinpoint the answer to his problem. He remembered her amused smile, confessing innocently with a slight shrug of her shoulders, that she had no clue what he was talking about.

Jacob's clothes were awry, one shirt corner sticking out of his pants, the belt undone, the knotted ritual tassels of his undergarment *tallith* hanging down to his knees. He washed his hands and face in the kitchen sink, then walked about to stay alert. Touching the *mezuza* on the doorpost, he stepped out of the apartment into the corridor where he recited the preliminary morning prayers from memory. The outline of a swastika, that had been spray painted on his front door a year ago, had faded but was still visible. Sarah had scrubbed most of it off with a soapy face towel. As he paced back and forth, a whiff of bacon slithered out of an apartment, reminding Jacob of the singed chicken feathers in the Samarkand markets of his childhood.

As Jacob prepared to go to the synagogue, Sarah stepped in front of him with an appeal: "Please stay home. Elka told me that no one wears masks in the synagogue. People have died in our community and many are in the hospital, especially the elderly. God will surely accept your prayers from home."

Jacob recalled the report from a New York website he had read in the night: a young Chassid, father of 6 children, had died from the corona virus in Boro Park. Heart rending scenes from the funeral made Jacob close the site.

At length, he said: "I'll stay home."

Sarah thanked him, then continued with the Sabbath preparations. She removed the tureen from the top of the fridge and placed it on the kitchen table to be polished. Jacob wondered if she had a prescient sense of what he was plotting to do with it and was subtly conveying to him the significance of the last remaining souvenir of their former life.

Jacob stood in front of his computer, wearing his phylacteries and draped in his *tallith*. He was holding a prayer book in his hands about to recite the morning service. Looking down, he realized he was holding a mathematics book. Horrified, he let the book fall to the floor. How could he have confused a prayer book with a mathematics book? No longer able to concentrate, he was now mumbling prayers by rote in a perfunctory manner. He surmised that instead of spiritually uplifting him, the algorithm had diverted him from the service of the heart. Saddened, he spoke to God: "Dear God, I, Your old data miner, ask for only a crumb of Your logic, visible through the veil of the HD numbers on my screen, a peep from behind the curtain, for a Planck-time, no longer. I implore You, at the very least, please traverse my screen as an amorphous digital mist and make one - only one - of your Biblical sequences match the DNA sequence of X. I have crafted and prepared it all so carefully for You. Then I'll return to my former self and serve You the way a Jew is supposed to."

． ． ．

Mid morning, a storm advanced on the city with military precision. First scattered flurries fell, then a steadily increasing snow fall, followed by a salvo of heavy wet snow. In three hours, more than six inches of snow had fallen. The seagull was covered in snow. Jacob went out to the porch and tried to throw bread crumbs onto the window sill, but the wind blew them into the alley below. He opened a space in the snow with his shoes and left bread for the seagull.

In the early afternoon, Sarah began cooking and baking *challah* for the Sabbath. Jacob paced the house nervously. From time to time, he lingered near the porch door. A strip of bark hung from a tree branch resembling a man's necktie. Jacob imagined that a Jew had been lynched on the tree a century ago. He thought, 'When the Messiah will arrive, that man will be resurrected from the DNA left on the neck of that tie.'

Jacob's thoughts shifted to the years before he had coded his algorithm, when he had contemplated an intermediate project; a preliminary step between the material world and the spirituality of Godliness. String Theory, with its higher, but still finite dimensions, is a first step above the world experienced with our coarse senses, striving to unite Quantum Mechanics and General Relativity in a super theory where the smallest pieces of matter and energy are tiny strings. These strings vibrate and the different types of vibration produce all the fundamental particles and physical forces. One of the requirements of String Theory is that at each point of space, there is a tiny 6 dimensional Calabi-Yau manifold. The Gordian knot is child's play compared to the shapes in six dimensions wedged between the material and spiritual worlds where Jacob hoped to glimpse traces of the supernatural. He imagined himself living in this micro Lilliputian land, with its interconnected atriums and ventricles, careening hairpin turns and steep valleys curved by extreme gradients. In this intermediate jungle, Jacob - Tarzan of the Calabi-Yau manifold - soared from submanifold to submanifold along catenaries of his imagination.

· · ·

So confident was Jacob of the Messiah's impending emergence from his algorithm, that he committed much thought to practical implications of the resurrection. What intrigued him most was the problem of past generations coexisting. If people who had died were to be resurrected as they lived in their respective generations, then there would be a jumble of generations on their resurrection. A man who died in the 19th century would be resurrected at the same time as his great grandson in the 20th century. It was obvious to Jacob that a consistent theory of resurrection required a new definition of time, a multi-dimensional landscape beyond human comprehension. Jacob had made several attempts to model resurrection mathematically. The most recent attempt imagined time as a higher dimensional variable where at every point of time, there existed a Calabi-Yau-like time manifold on which past generations were brought to life as they lived. On this manifold, there was sufficient room for all previous generations to live together at the same time.

With only minutes before the Sabbath began, Jacob took one last glimpse at the latest HD results. There was the usual range of random numbers on display. In a moment he would step away from the computer and ten minutes later, the screen saver would turn the screen black, but the program would continue running in the background, inaccessible to Jacob for the next 25 hours.

He dropped several coins into the charity box while Sarah lit 3 candles on the kitchen table: one for Jacob, one for Elka, and the third for herself. The mellow sound of her benediction lifted him up out of his stressful week on eagle's wings and lowered him gently on an oasis in the desert of time, the holy Sabbath. Jacob responded *amen* to her blessing.

The evening brought calm. With the computer screen off and the HD results completely out of sight, Jacob was at ease. Sarah served him two slices of fish. He pretended he was full and placed one slice back on her plate.

"Eat," she insisted.

"I'm full," he said, feigning satiety. Sarah wrapped the slice in aluminum foil, then put it into the refrigerator. Jacob ate a chicken leg, but Sarah saved her portion for a grandchild should one of them drop by Saturday afternoon. Jacob had a habit of picking up the crumbs of *challah* crust on the table with his dampened forefinger; it reminded him of the hunger he had endured as a child in Russia. After they recited the concluding blessings for the meal, Sarah went to bed but could not sleep.

She lay awake thinking about an evening at the country cottage when Elka was a toddler. The setting sun flashed through the forest canopy behind the cottage. From the open patio door, the braying of sheep from a nearby farm wafted in with the breeze. Sarah had tiptoed into Elka's room. Elka was lying on her back, her left arm curled over her head, one heel touching the other knee, a ballerina in sleep. The sight of a peacefully sleeping child intimated terror on Sarah's face. She had spent her nights at Elka's bedside, crouched, prepared for combat with Lilith, never doubting her appearance. A wedge of light from the staircase revealed the toddler's face, mucous crusting her nostrils. Elka had a cold; Sarah regretted having allowed her to stay so long at the windy beach that morning. She stretched the tip of the pink sleeper over the exposed toes, then bent forward to listen for the child's breath as she had listened to her mother's breathing when she was a child. Sarah pressed her trembling hand against Elka's chest and waited for a force within her to react against it. At last it came, a magnificent expansion of the tiny lungs, and Sarah sighed with relief as though there had been a real doubt. Sensing Sarah's presence, Elka woke up. She needed a snack. Seated in the rocking chair, Sarah unbuttoned her blouse. Eyes closed, Elka's open mouth searched in the half-darkness. Many months after she had planned to stop breastfeeding, Sarah was still nursing, clinging to the last vestige of youth. Elka's hair was sticky with sweat, her head heavy in the crook of Sarah's arm. Suckling rhythmically at the flaccid breast, the toddler slipped back into sleep. Caressing Elka's fat little hands, Sarah hummed Yiddish tunes she had heard as a child in the DP camps of post war Europe. Sarah's probing fingers revealed a soaking diaper. As she changed Elka on her lap in the glow of the staircase light, she noticed the delicate appearance of a new tooth, a little wound filled with ivory, pushing up at a slight angle to the curving ridge of the pink gum. Sarah clicked it lightly with her fingernail, delighting in the sound. With the child's head pressed tightly to her shoulder, Sarah lowered Elka into the crib on her back. Feeling the absence of mother's touch, Elka's arms reached up stiffly in the darkness, fingers outstretched. Into the child's gesture, Sarah read a painful truth about herself, about the father she had lost in the ocean, whom she failed to save with her outstretched arms. Sarah had an intense desire to kiss Elka on the lips, but she refrained, bundling the warm entropy of the aborted kiss into the resolve to stabilize her life now that she was a mother, not a mother of many as she had hoped, but nonetheless a real mother.

"I love you, my precious. I love you," she whispered, her eyes glazed with tears of bliss and sorrow.

CHAPTER
ELEVEN

JACOB STEPPED QUIETLY into the doorway of the bedroom. Sarah was weeping in the darkness. He sat down on the edge of the bed, took her hand in his, and caressed it until she fell asleep. While listening to her breathing, he dozed off and dreamt of the Rebbe's funeral to which he had taken 13 year old Elka. Eastern Parkway was restricted to pedestrians only, with thousands of men, women and children gathered in front of 770. At 4 PM, the coffin emerged. Atop the casket, the Rebbe's black hat bobbed up and down, a black horse without a rider. The coffin was pushed into the hearse, which moved slowly through the crowd. Jacob mourned the revered man who could recognize the divine spark in every Gentile and Jew. Elka wept with him. He held her closely, contemplating how the orphaned world would cope without the Rebbe's guidance.

Jacob stood up from the bed and made his way to the sofa. The memory that had escaped him when he faced the cashier in the pharmacy suddenly flooded back, drenching him in sweat. The blonde hair and the sound of the cashier's voice had stayed in his thoughts and now, two days later, evoked the defining incident of Jacob's teaching career at the university. It occurred at the end of the fall semester of 2016 when Jacob had taught a course on Differential Equations. Among the students was Donna Walters, a stout, blonde woman in her late thirties. From the first lecture, Jacob recognized that Donna was pretending to grasp the material, nodding excessively as if effortlessly comprehending everything. After class, she approached Jacob with questions that made it clear she did not belong in this course. She confided in Jacob about her plans to pursue computer science and her absolute need for an A to gain admission. Such warnings were not new to Jacob. As he expected, Donna failed the final exam in December, receiving 37/100 after Jacob's generous grading. She had not taken the proctored midterm, but submitted flawless assignments in hand-writing markedly different from the incomprehensible scrawl on her final exam. Donna requested a meeting to review her exam.

"Good morning, sir," she greeted Jacob, standing stoutly in the open doorway, her smile unwavering. Jacob gestured for her to come in, and she took a seat across his desk. Jacob painstakingly pointed out her errors, providing the correct solutions. She contested each mark deduction, reluctantly acknowledging blatant mistakes. After two hours of back-and-forth combat, she pushed her chair back, placing her booklet on her thighs to scrutinize problem 10 on the last page. She had started it incorrectly and left it unfinished, earning 0 for the problem. Jacob noticed her using her smartphone, and shortly thereafter, the desk phone rang. Jacob turned aside to take the call. It turned out to be a wrong number. When he returned his attention to Donna, she was focused on the front page of her exam booklet, holding a pen in hand. Later he would realize that the phone call was timed to distract him from the student while she wrote something on the front page of the booklet. The meeting concluded with Jacob standing firm; he would not allow a make-up examination which would enable her to earn a higher grade.

A few days later, Donna applied for permission to write a new examination based on the argument that her second booklet was missing and that it had the correct solution to Problem 10 which was started on the last page of booklet #1. She wrote to the chair of the department, alleging that Professor Lazerson had admitted in their meeting that a booklet was missing and promised to have the administration order a new examination for her. She presented a recording of Jacob stating this and, as further evidence, claimed the number 2 written on the top right corner of the front page of the first booklet indicated that two booklets had been submitted. Anticipating the question that the number 2 was written over the number 1, she argued that when the exam began she was confident that she would not need more than one booklet. However, towards the end she realized she needed another booklet so she went back to the front page of booklet #1 and overwrote the #1 it with 2. Jacob was convinced that this was done in the moment when he turned away to answer the phone call. He was absolutely certain that during their meeting there was no mention of a second booklet. A few days later, he learned that the student had taped their conversation and doctored the tape. At a meeting with the chair, the associate chair, and the executive secretary of the department, the student produced a tape with Jacob's voice that convinced the administrators that Jacob had in fact promised her a new examination based on the missing booklet. After the meeting to which he had not been invited, Jacob was called to the chair's office and presented with 'irrefutable' evidence of misconduct on his part.

"Jacob, you do not have the authority to promise a student a new examination." the chairman declared.

Jacob replied: "I did not promise the student could write another examination. What I said was that the paper she wrote could be reread by another professor."

"Yes, I understand, but that is not acceptable to her because booklet #2 was missing. She claims you acknowledged this fact."

Jacob asked to hear the tape, but the request was rejected as the three departmental administrators were absolutely convinced by the voice evidence. Gazing with disappointment at the chairman, Jacob said, "In the forty years we have known each other, have I ever lied to you?"

"No, of course not, Jacob," the chairman replied matter-of-factly. "I never suggested you did. But we're all getting older, prone to forgetfulness. We say things we do not remember. Now that you promised the student a new exam, you will have to keep that promise. But another professor will grade the exam." Deeply disheartened, Jacob knew that his days at the university were numbered.

A year later, after dedicating fifty years to the university, he resigned. A young Chinese professor specializing in Algebraic Number Theory was given his office. In recognition of Jacob's long service, the university gave him a leather briefcase and allowed him to retain his email address for a period of 5 years.

The apartment was cold. As it is prohibited to adjust the thermostats on the Sabbath, Jacob sat down on the sofa and drew a blanket over him. In the dim light from the kitchen, he picked up the Klondike book. Skimming through the pages, he paused at photographs taken at Sheep Camp after the avalanche. Jacob had survived and pressed on to the cleft in the mountain peaks from where remained a four hundred mile journey down the Yukon River. There he met Freddy Rapoport a few days before he was swept off a barge and drowned in the White Horse Rapids. In Dawson City, Jacob walked through the muddy streets lined with raw sewage, sporting his skull cap, a shovel tied to his shoulder, and a hammer interlaced with one of the 4 long fringes dangling from the *tallith* under his shirt. Music blared from a tavern. Jacob staked his claims to spiritual gold which he hoped to find in the permafrost, the sequence of Hebrew letters matching the DNA sequence in X. In a saloon, Jacob heard the world famous atheist, Dr. Richard Dickens, who recently arrived from London, lecturing on evolution. It was now Jacob's turn to address the rowdy crowd. Standing on the saloon counter, he spoke in a loud voice: 'We are here to mine for gold, but we must know that God did not create gold for us to become rich. Gold was created only for beautifying the Holy Temple in Jerusalem. Because mankind was granted free choice, that gold was sprinkled throughout the earth in a manner that it can be found by anyone and used for purposes other than the Holy Temple.' A prospector threw a raw egg at Jacob as he jumped down from the counter and went to a roulette table. Jacob was now a riverboat gambler. Rolling the dice, he screamed: 'Come on, Mr. Hamming, give me the big zero to bring the Messiah!'

· · ·

Saturday morning, Sarah found Jacob in his clothes, asleep, leaning into the corner of the sofa with the Klondike book face down on his chest.

"Jacob, Jacob, you could have slept in a bed like a normal person," she admonished him disappointedly. "Now your head will be foggy all day."

Jacob could not reply. After drinking a cup of coffee and eating a few almond cookies, he slipped his shoes into his boots, then twirled his prayer shawl over his head, allowing it to settle on his shoulders. Sarah helped him don his winter coat. The fringes of the prayer shawl hung below the coat, barely above the ground. Sarah escorted him to the staircase. The mask had slipped under his contused, aching nose. Sarah adjusted it and told him to wear it in the synagogue even if congregants ridiculed him.

Jacob immersed himself in the ritual bath in the synagogue basement. Men entered and exited the bath as Jacob stood in a corner, the water up to his chin, eyes closed. Though it wasn't appropriate to linger there on the Sabbath, the steam made him too weak to ascend the seven steps to the floor. He was thinking about the grocery store where he had been tempted to buy grapes, and very nearly reached out to grab a long bunch that he recalled was shaped like the coastline of Italy. If his algorithm had actually refined his baser instincts, he would not have been tempted to reach out for the grapes. Had he not learned many years ago that contemplation of sin is worse than sin? Contrary to his hope at the outset of his great experiment, the algorithm had failed to rein in worldly temptations. How presumptuous of him to have thought that a few hundred lines of Java code could elevate him to sainthood, and simultaneously entice God into a scientific test. Instead, the algorithm had become a reptile in his hand, an image used in the Talmud for spiritual defilement, invalidating the immersion.

Jacob was one of the few congregants wearing a mask in the men's section of the large Sabbath synagogue that served as a school auditorium during the week. While many may have understood the risks of contracting the corona virus, the majority did not wear masks, fearing it might intimate that their faith in God to protect them was lacking. During the service, Jacob sat at the back of the synagogue, his prayer shawl covering the back of his head. He recited prayers from memory, not with the passionate intensity of a devout Chossid. The cantorial embellishments of the service irked him. Drifting into a light slumber, his skullcap slipped off his pate to the floor. He found it beneath a wooden bench. The skullcap reminded him of the time when Quebec's government prohibited religious symbols in public sectors. Female professors could not wear headscarves, and male professors were not permitted to wear skullcaps. Yet, crosses were permitted, seen as representing the province's cultural heritage. Long ago, he had grown accustomed to the bigoted logic of the government.

Jacob loved engaging with the children, handing out lollipops. Once they had tasted the sweets, he playfully asked for them back, promising even bigger ones the next day. Enjoying their treats, none took up his offer. Through this, Jacob hoped to discern evidence of a unique trait in their DNA - one that symbolized trust and the capacity for delaying gratification - qualities he believed that had upheld the Jewish people's faith as they patiently waited millennia for the Messiah.

To Jacob's surprise, he was beckoned to the podium to lift the Torah scroll. Grasping its wooden handles, he attempted to raise it. As he did, the handle in his left hand quivered, causing the scroll to teeter. A collective gasp filled the synagogue. Jacob promptly stabilized his hold, lifting the scroll triumphantly. He rotated the scroll to the right and then to the left before gently setting it back onto the dais. The congregants exhaled in unison with relief; if the scroll had hit the ground, they would've been obligated to fast for 40 days.

During the typically long sermon, Shmuki took a seat next to Jacob. Draping an arm around Jacob, he whispered Psalms. Following the morning services, a *farbrengen* took place. The men sang Chassidic melodies and delved into religious discussions while partaking of gefilte fish and cakes, and sipping liquor. The melodies reminded Jacob of the times when this place echoed with tales of the great chassidim, whose hardships and martyrdom in the Ukraine and Russia had inspired the younger generation.

Shmuki saw Jacob attempting to surreptitiously leave the synagogue. He approached Jacob, took his hand, and guided him back to the *farbrengen*. Within minutes, Shmuki was becoming inebriated. He climbed onto a bench, wrapped in his prayer shawl, embracing and kissing anyone who came close enough.

Jacob remembered a weekday Purim celebration 35 years prior. Some men had lingered late into the night with the Rav, the community's spiritual leader. One of them had tricked the Rav's wife into giving him a blank sheet of the Rav's letterhead. The Rav rarely drank, but today he found himself tipsy, mingling and chatting jovially with men he would typically avoid. Seizing the moment, Shmuki convinced the Rav to draft a declaration urging the Rebbe to proclaim himself as the Messiah. When it came to sign the document, the Rav hesitated. Shmuki offered him a glass of 96% Polish vodka. The Rav swallowed it in a single gulp, then proudly signed the document. Moved to tears, he expressed regret that this letter had not been presented to him earlier; the two thousand year Diaspora would have ended long ago and the Jewish people would now be with the Messiah in Israel.

Days later, the community learned that the Rebbe had disapproved of the Rav's letter and reproved him sharply. For months, the Rav walked around with a lowered head.

Jacob came home 2 o'clock in the afternoon. After a modest meal of fish and *challah*, Sarah tidied the kitchen, then carried on with her Psalms. Jacob had a tradition of visiting people in the hospital every Sabbath afternoon. Even now, when hospitals were closed to visitors, he felt it was his duty to try to get in. Denied entry at the main entrance on Cote St. Catherine, he tried the Legarie Street side door which was also locked. Tired, he rested on a wooden bench in the vestibule between the exterior facade and the interior glass doors and gazed down the long corridor toward the brith room. He found himself reminiscing about an incident with Shmuki. While studying at the yeshiva, Shmuki also worked part-time as an assistant food inspector at the Jewish General Hospital. In the hospital kitchen, he overheard someone say that an acquaintance of a friend of Shmuki's had a third cousin who had given birth to a boy in this hospital. The baby was scheduled to be circumcised by a Gentile plastic surgeon. When Shmuki heard this, he took matters into his own hands. He had decided that the baby must have a kosher circumcision. Having circumcised many babies, especially in distant towns where the established *mohels* were reluctant to travel, he was considered to be an expert in the community. On the morning of the *brith*, Jacob met Shmuki in the main lobby of the hospital. Having access to hospital scheduling information, Shmuki knew the surgeon's name and office number. He waited for the surgeon to leave his office, then followed him down 6 flights of stairs to the basement level. As they approached the landing, Shmuki caught up to the surgeon and from 3 steps above leaped on him, covering his nose with an anaesthetized gauze pad.

"Inhale!" Shmuki commanded. The surgeon complied and in a few seconds he was fast asleep. Jacob had been waiting on the landing. They seated the surgeon on the floor with his legs stretched out toward the door to block intruders. They removed the doctor's white coat. Shmuki quickly slipped into it, then rolled up the sleeves. Dressed in full surgical regalia, with a blue crepe hat, gloves, his beard tucked tightly into a mask, Shmuki glanced at the hospital coat and read the name tag, Dr. Ian McClintock. In the *brith* room, Shmuki introduced himself as Dr. McClintock with his Yiddish accent. Glancing around the room, he recognized holocaust survivors and Soviet Jews who had recently immigrated to Canada. Jacob whispered anxiously to Shmuki: "You're taking too long! The doctor will wake up any second. Cut it off already!" Shmuki dipped the knife in alcohol. Then, swiftly and skilfully carried out the task while Jacob stood guard in the corridor. Biting on a wine-dipped gauze pad, the baby was brought to his mother to nurse. The father slipped five one hundred dollar bills into Shmuki's hand, but he refused payment. Shmuki did not want the gift to detract from the fulfillment of God's precious commandment.

Someone was shouting in the corridor. Jacob ran into the *brith* room and tugged on Shmuki's sleeve.

"We have to get out of here fast!" he whispered into Shmuki's ear.

Feigning calm, Shmuki nodded politely to family and their friends, and once in the corridor, sprinted toward the Legarie Street exit with

Jacob on his heels. Someone was chasing them and in front of them, a tall guard was blocking their way. He grabbed Shmuki by the back of his collar and lifted him up. Thrashing his arms and legs, the guard lowered Shmuki to the ground. Fortunately, Shmuki had taken boxing lessons as a teenager and knew exactly what to do in circumstances such as this. He aimed a kangaroo 1-2 punch below the guard's belt, followed by a swift upper cut to the chin that forced the guard to slump backwards. Shmuki deftly pulled up a chair behind him and guided him into the seat. With his thighs tightly together, mouth agape, eyes crossed, the guard cursed indistinctly in Quebecois French. Once on the street, Shmuki realized he was still wearing the surgeon's jacket. He pulled it off and threw it on the shoulder of a man heading into the hospital. Shmuki did not want to be accused of theft.

Shmuki's audacious behaviour inspired Jacob. It gave Jacob the push he needed to tackle his mathematical problems.

Denied entry, Jacob left the hospital grounds. When he arrived home, he found Elka's seventeen-year-old son, Shloimy, sitting with Sarah in the kitchen. Jacob playfully rubbed the stubble on Shloimy's face with the back of his hand. Sarah gave Shloimy the chicken she had leftover from Friday night, chatting with him as Jacob delved into the Bible at his desk. Occasionally, Jacob would glance at his computer's dark screen, hoping for a notification of success that was programmed to break through the screen saver when that moment arrived. Some rabbinic teachings suggest the Messiah will in fact arrive on the Sabbath. Holding onto this belief, Jacob imagined the Messiah's arrival in the form of a plain notification on his screen displaying the Biblical sequence that perfectly matched the sequence of bases in X. There would be no preliminary drumbeat rising to a crescendo. The Messiah would reveal himself humbly on the screen, a man dressed in ordinary clothes, a broadcaster with true information to disseminate to the world.

. . .

Shloimy coughed frequently. Holding a glass of orange juice, he approached Jacob but stumbled on a corner of a slightly raised tile, causing the juice to splash onto the keyboard. Although Jacob bolted to his feet, he could never show annoyance towards Shloimy. How fondly he remembered their hide-and-seek games when Shloimy was a child. Once Jacob hid so well that the child could not find him and burst into tears. When Jacob revealed himself from behind the living room curtain, he was met with Shloimy's joyful tears. Jacob also wept, wishing he could seek God with the pure feelings of a child.

Shloimy found a towel and tried soaking up the juice from the desk.

"Sorry," he said, his face red with embarrassment.

"Don't worry," Jacob reassured him. "This has happened before. The keyboard will dry and be fine. Let's continue our Talmudic studies," he added, hoping to divert Shloimy's attention from the keyboard. "Where did we leave off last week?"

"Berachot, page 18," Shloimy replied.

Jacob took the large volume of the Babylonian Talmud from a stack of biblical books on his desk. He had prepared this page for his grandson weeks before, consulting numerous sources to clarify its complexities. Flipping to page 18, Jacob translated as he scanned the Aramaic text: "There was a story of a pious man who gave a coin to a pauper during a famine year. His wife scolded him, so he went to sleep in the cemetery where he overheard two spirits talking to each other. One said to the other: 'Let us go travel in the world and hear from behind the curtain what punishment is coming to the world.' Her friend said to her: 'I cannot because I am buried in a mat of reeds. You go and tell me what you hear.' She went out and came back. Her friend asked: 'What did you hear behind the curtain?' She said to her: 'I heard that anyone who plants at the time of the first rainfall will have their crops struck by hail.' The pious fellow went and planted during the second rainfall. The entire world's crops were stricken but not those of this fellow. The next year, he went to sleep in the cemetery. He heard two spirits talking to each other. One said to the other: 'Let us go travel in the world and hear from behind the curtain what punishment is coming to the world.' Her friend said to her: 'I cannot because I am buried in a mat of reeds. You go and tell me what you hear.' She went out and came back. Her friend inquired: 'What did you hear behind the curtain?' She said to her: 'I heard that anyone who plants at the time of the second rainfall will have their crops struck by blight.' The pious fellow went and planted at the time of the first rainfall. The entire world's crops were struck by blight but not those of this fellow.

His wife asked him: 'How is that last year everyone's crops were stricken except yours, and now everyone's crops were blighted except yours?' He told her the whole story. 'It was not many days before a quarrel developed between the pious fellow's wife and the mother of that deceased girl. The wife said to the mother: 'Come and I will show you your daughter buried in a mat of reeds.' The next year, the man went to sleep in the cemetery. He heard those spirits talking to each other. One said to the other: 'Let us go travel in the world and hear from behind the curtain what punishment is coming to the world.' She said to her: 'Leave me be. The words passed between us are heard among the living.'

Jacob said to Shloimy: "We see that the deceased know what happens in this world. The man acted on this knowledge, saving his crops twice. But when his wife shared the spirits' secret, they became wary of speaking again."

"What a strange story! Is it true?"

"I don't know," Jacob replied.

"Zeidi, I'm not feeling well," Shloimy said, grimacing in discomfort. "I better go home."

As Shloimy prepared to leave, Sarah embraced him, then felt his forehead with her lips. It was warm. Jacob escorted Shloimy to the front door of the building and watched him walk down the street. Once back in his apartment, Jacob ventured close to his desk and stood behind his chair gazing at the black screen. Inadvertently, he pushed the chair forward; it struck the desk causing the mouse to move, which activated the screen. Jacob was grieved by his unintentional desecration of the Sabbath. How could he have allowed himself to come so close to the desk! Had he forgotten that in Jewish Law accidental was tantamount to intentional?

· · ·

As dusk settled on the quiet streets, Jacob walked back to the synagogue. Between the afternoon and evening services, the men gathered around the long tables where the afternoon *farbrengen* took place, singing melodies that expressed a yearning for the divine. As a toddler, Elka used to sit on Jacob's lap at these tables. He would gently blow on her back, and she would shrug her shoulders, laughing. How he had wished he could capture those moments in a universal theorem.

In the evening darkness, Jacob tried to hold onto the serenity of the Sabbath. As he approached his apartment, he heard Sarah moving inside. Entering, he noticed she had prepared the wine, candle, and incense for the *Havdalah* prayer, marking the Sabbath's end.

After *Havdalah*, he doffed his black caftan and quickly turned to his computer to review the HD numbers generated during the Sabbath. Waking the computer with a mouse click, he noted the keyboard felt sticky but functioned. Jacob scrolled up the left window to see that HD had made a few moves towards 0 but nothing of any significance could be deduced from the data. Jacob studied the chaotic HD results and felt like scolding the algorithm, 'How could any self-respecting algorithm produce such meaningless sequences of numbers?'

Feeling deflated, Jacob switched to the latest news about the pandemic in New York. Several renowned Torah scholars had passed away. The number of dead in New York City were alarming. Many were avoidable consequences of major transmission events.

The phone rang. It was Elka, informing her parents that Shloimy had tested positive for the corona virus. She had been advised not to take him to the hospital as young patients were being turned away. Sarah immediately penned a letter to the Rebbe, which Jacob faxed to the Rebbe's gravesite. They then recited Psalms for their grandson. Sarah's tears flowed as she whispered the words. Wanting to reassure her, Jacob mentioned articles he had read on the internet. Most young people experienced mild symptoms, and a Chassidic doctor in New York had seen success using Hydroxychloroquine. But the drug wasn't available in Canada. Jacob knew someone who had gone to Plattsburgh, New York, a few miles past the Canadian border, for the drug.

However, only an American doctor could prescribe it. Jacob reached out to Dr. Rosenberg in New York, a contact of his, and secured a prescription which the doctor sent to a Plattsburgh pharmacy. The challenge now was to get there before an expected border closure. Jacob still had his driver's licence but they had to rent a car. All the local car rentals were closed. Only the airport rental places were open. Jacob and Sarah hurried to the Van Horne shopping mall and took a taxi to the airport where they rented a car. At the border, the guard questioned them: "Any symptoms?"

"No, Sir," Jacob replied. Beside him, Sarah focused on her Psalms.

The medicine was ready when they arrived at the pharmacy. By 2 AM, Jacob and Sarah were back in Montreal. They brought the medication to Elka's apartment, handing it to Chaya, Elka's daughter. Then they parked the car at a Decarie lot and walked home in the silent darkness.

CHAPTER
TWELVE

A LITTLE MORE AT EASE, Sarah prepared a cup of tea for Jacob, then went to sleep. Jacob returned to his desk and pushed up the sleeves of his Sabbath shirt, gearing up for the few hours left in the night. As dawn approached, Jacob rested his head on his desk and dreamt of traveling to the Paranal Observatory in the Atacama Desert of Chile to meet Dr. Richard Dickens, the world's leading atheist. Jacob ascended the mountain on foot, while Dr. Dickens arrived by helicopter. They met and exchanged pleasantries atop the plateau. The night sky above them was dotted closely with stars.

"Dear Dr. Dickie Dickens, please forgive the alliteration. It certainly was not intended. English grammar was never my strength. It's an honour to finally meet in person the pope of atheism."

"The honour is mutual, dear Dr. David Lazerson."

"Jacob. Jacob Lazerson."

"I stand corrected."

"Dr. Dickens, I recall your conversation in Melbourne with your Jewish colleague – the name slips my mind."

"Dr. Hymie Kriss. Melbourne 2010. Recorded on Youtube."

"Ah, yes, Dr. Kriss. I researched him after watching the video. His original family name was Kresovsky, from Otvosk, Poland, a two hour horse ride from Warsaw. From the intellectual exchange between the two of you, you impressed me as someone who would have risked his life to save Jews during the Holocaust. You would have certainly been inducted into the Yad VaShem."

"I appreciate that."

"But, I must admit, I was taken aback by Dr. Kriss's disrespect towards believers, especially the Archbishop of Canterbury, who is a dear friend of mine. Please let me ask you, why do you think it is important that an Archbishop be as versed in evolution as Dr. Kriss? Does Dr. Kriss know any details about the Spanish Inquisition of 1478? At any rate, this is not the reason I have invited you here. We are here in the hope we can bridge the gap between our beliefs. First, I would like you to look at the magnificent sky, filled with millions of galaxies of the Milky Way. Do you see Canis Major there? Yes, just beneath that hazy band. Dr. Dickens, in simple words, what do you perceive?"

"A breathtaking tableau. Truly awe-inspiring. Such brilliance! And what about you, Dr. Lazerson?"

"I see the light that transcends the light you see. I see the Creator of all that we see."

"I see the speck of ultra compressed hot gas from which all this evolved."

"A speck became the universe?"

"Yes."

"Where did that speck come from?"

"From another universe."

"Dear Dr. Dickens, with all due respect, that is hardly a satisfactory answer."

"Scientists focus on physics, not metaphysics, Dr. Lazerson. Our truths come from mathematical equations."

"Dear Dickie, I feel we are so close: you state the universe began with a speck from another universe and that universe came from a speck of an earlier universe in an endless backward succession, and I claim the universe started once and from nothing. The difference between us is the difference between an infinitesimally small speck and nothingness."

· · ·

The whirring sound of the computer's hard disk brought Jacob back to reality. The screen was filled with white lines. Despite reboot attempts, the computer remained unresponsive. Luckily, he had a backup that saved his data every hour on a USB drive. Another USB stored the latest versions of the packing and comparison programs. After unplugging the computer, he opened the side panel and removed the hard disk that had served him well for 10 years. Jacob had prepared for this moment and knew exactly what he wanted to buy. With a dull pain in his chest, he knew that he would have to pawn Sarah's tureen.

Sunday morning, Jacob stood at the porch door, peeking between buildings to the park where children frolicked in the snow. He observed a section of an ice rink, with skaters passing in and out of the view of his single slit experiment. Elka phoned to update her mother about Shloimy's deteriorating health. *Hatzolah*, the community volunteer medical service, had assessed his symptoms which matched those of the corona virus. Being young, they hoped he would recover soon. As Jacob recited Psalms for his ailing grandson, he felt the weight of sharing with Sarah the need for a new hard disk, given the work he had invested in his project. However, he couldn't bring himself to disclose his plan for the tureen.

As he was about to leave, Sarah reminded him, "Wear your mask." Jacob had secretly tucked away the tureen under his coat.

The Sikh owner greeted Jacob with an affable smile.

"I hope to reclaim it soon," Jacob said as he handed over the tureen.

The owner gave $250 to Jacob, who felt a deep sadness at what he had to do in order to proceed with the project that had taken possession of him. Exiting the shop, he was met with a bleak sky uniformly the colour of fingernails. Snow clung to power lines, and melting icicles occasionally dropped to the sidewalk in the warming weather. Muddy puddles at street corners brought back memories of a Sabbath morning in Samarkand. Jacob's mother had prepared him for what she was about to do. She sat him down in a muddy slough and smeared mud over his jacket and pants. With dirty clothing, she had an excuse to tell the teacher that she had to keep Jacob home.

Jacob found himself among South Asian pedestrians, who were occasionally hidden by snow banks. A child's shoe on a pile of soiled snow summoned wistful memories of shtetl life: children running through dusty streets on a warm summer day. Nearby, his attention was drawn to a discarded cigarette butt smeared red with the DNA of a young woman, imbedded in ice for a future generation to discover. On reaching Jean Talon Avenue, he made his way to the Best Buy Store. Jacob wanted to buy a high-speed disk, knowing the improvement it would bring to his program's performance and thereby accelerate the appearance of the Messiah. A Muslim woman assisted him at the computer counter, suggesting a better model than the one Jacob wanted. It had been on a sale that ended yesterday, but which she would extend for him. Jacob thanked her, thinking that such an act of kindness between different faiths had the power to uplift spirits across the world.

Jacob approached the usually bustling Decarie Expressway, now eerily silent. A massive construction site on the western side stood still, the pandemic having halted its progress. Partially built structures and pillars hinted at the promise of what was to come. Conscious of the approaching afternoon service, Jacob paused at a tree. Gazing upward, he saw the half white evening moon snagged in the tree's branches. It said, 'Good afternoon, Professor Lazerson.' He acknowledged the greeting with a polite nod, then facing east towards Jerusalem, he recited the afternoon service from memory.

The algorithm's installation on the new hard drive proceeded smoothly. Nonetheless, many hours had been lost. To recover lost time, Jacob decided to forego his Torah study Sunday evening. Furthermore, darkness from the outside inundated the rooms with their meagre lighting and made reading difficult. Noticing Sarah sitting solemnly at the kitchen table, he presumed she had already become aware of the tureen's absence. Overshadowed by the enormity of his goal, the theft did not trouble him as much as he had feared. He even justified his deed by reminding himself that they had not used the tureen for more than twenty years and it was doubtful they would ever need it again.

Later that evening, Sarah began to feel chest discomfort and feverish. Persistent harsh coughs produced green phlegm. He felt her forehead. It was warm. She was shivering and her teeth chattered uncontrollably. He gave her tea and two Tylenol tablets. Then he went to his desk to draft a letter to the Rebbe, informing him of Sarah's health and requesting a blessing for her recovery. When he returned to the bedroom, Sarah was lying on her side, breathing with difficulty.

"I'm fine," she whispered hoarsely to allay his fears. "It's just a cold."

Jacob felt certain that he was being punished for having ventured too close to life's source. He knew the rabbinic view that the Messiah's coming should not be the result of human effort. But he did not subscribe to this opinion even as Sarah's symptom's seemed to worsen; he had invested too much in his program. Earlier in the night, he had heard a knock on the front door. Now he understood that it was the corona virus announcing its intent to infiltrate their home.

With her chest heaving, mouth open, Sarah lay on her back. Jacob shifted her sweaty head to a drier part of the pillow, then called *Hatzolah*. Fifteen minutes later, three men wearing protective gear, entered. One held a digital thermometer, another a pulse oximeter, and the third was ready to call 911 if necessary. They checked Sarah's vitals: her temperature was elevated and her oxygen levels dangerously low.

"She needs high flow oxygen," one of them affirmed.

They had an apparatus in one of the cars. Two of the men retrieved it, but when powered on, it did not work. After a brief consultation, they told Jacob that Sarah needed to be hospitalized for proper oxygen treatment. Ten minutes later, an ambulance arrived. Sarah was carried down the stairs in a stretcher, enveloped in a grey woollen blanket. Following closely, Jacob noticed the psychedelic ambulance lights painting a sinister scene as Sarah was lifted into the ambulance. She called to Jacob who stood at the back doors in the cold, wearing only his white shirt, that she had forgotten her Psalms on the bed stand. Jacob hurried back to get it. As he returned it to her, she said raspingly, "You didn't eat your porridge this morning." Jacob handed the Psalms to her. "You have to eat," she pleaded. "Please, Jacob, take care of yourself and take care of Elka." Jacob attempted to climb into the ambulance, but was restrained by one of the paramedics. The doors were locked and the ambulance drove off, slowly ascending the slippery hill toward the hospital two blocks away. Jacob ran after the ambulance along the sidewalk. A high mound of snow blocked his way. He slipped between cars and took to the road which had been partially ploughed. He fell into a snow bank and lay there. Behind him a car stopped, its lights glaring. A French-speaking woman, wearing a black mask, bent down beside him in the snow, asked him how he was, then helped him to his feet. He convinced her that he was well and thanked her with a nod and a 'merci.'

At the electric sliding doors to the ER, Jacob was refused admission by a guard who told him that a nurse would come to see him. Jacob paced back and forth in the vestibule between the doors. At last, a nurse approached Jacob. He stood up.

"Your wife has been intubated and her vital signs are stable. You're best off to go home. We have your home number. Someone will call you soon."

"What does soon mean?" he asked impatiently.

"I don't know, sir," she replied.

Jacob stayed on a little while to recite Psalms from memory while sitting on a marble ledge, his back against a window overlooking a concrete courtyard.

Jacob then decided to pray the evening service with a quorum. There was a backyard synagogue near his apartment, one of several that had sprang up in the neighbourhood during the first months of the pandemic. As he made his way along the path girdling a duplex, he entered a makeshift tent where he was greeted warmly by men of his community. Together with Jacob, there were exactly ten men present, the quorum needed for communal prayer. Jacob was wearing a thin jacket. The owner of the backyard had an extra coat which he placed on Jacob's shoulders. Frozen rain pounded the tent, pouring noisily down the sides of the roof, turning the earth around the tent into mud. Jacob stood slouched, tears flowing as he prayed for Sarah, beseeching the Master of the Universe to have mercy on his saintly wife. When the service ended, the owner of the yard refused to take back his coat, insisting it was an old torn jacket he kept just for emergencies.

. . .

Sarah's black sweater lay on the living room floor, rolled up, one sleeve protruding like the tail of a dead cat. Jacob picked it up and dropped it on the sofa where he sat down to recite Psalms for his wife. In the middle of the night the telephone rang. A doctor from the ER informed Jacob that Sarah had been diagnosed with the corona virus. She had been connected to high flow oxygen and was prescribed dexamethasone. With an aching heart Jacob recited the entire Book of Psalms.

The following morning, Jacob was stopped by a security guard at the entrance to the ER. An hour later, a nurse came to tell him that during the night Sarah had been transferred to the K-pavilion corona virus ICU. Jacob knew his way around the hospital. He took a stairwell that was rarely used now. From there he crossed into the K-pavilion behind security lines. He then climbed the stairs to the third floor and found his way to the ICU where electronic doors barred his further advance. A telephone on the wall nearby was meant to reach the nursing station inside the ICU. Jacob tried repeatedly. Finally a nurse answered. She informed Jacob that Sarah was on a ventilator and asleep. She assured him that another nurse would call him in the evening.

On arriving home, Jacob felt weak. He pressed two fingers of his right hand against his left wrist and counted his pulse while following the seconds hand on the kitchen wall clock. The pulse rate was over 150 per minute. He remembered that he had forgotten to fill the prescription for his atrial fibrillations and had not taken those pills for weeks. The last appointment with the cardiologist came to mind. After listening to Jacob's heart, the cardiologist stepped back with indignation and disgust. "Next time, before you come here, use mouth wash!" he said angrily. Jacob turned away with the abject silence of old age. A Chassidic thought raised his spirits: being unjustly insulted yet joyously accepting the insult, God would surely balance the equation of fortune and compensate him with good news from the algorithm. The law of conservation of luck was no less a law of nature than the law of conservation of momentum, and so he expected to be soon rewarded with success, a Messiah-permeated Bible sequence perfectly matching X.

Jacob walked to the main synagogue to pray for Sarah. When he arrived he found the front door locked. He was late for the evening service but he wanted to go inside. There was a 3-number code for the combination lock which he could not recall. He worked with 959 length codes whose first hundred bases he knew by heart, wrote lengthy code in Java, but could not recall a 3 number code that would permit him entry into the synagogue. Jacob returned to his apartment and prayed for his wife and grandson. To advocate on their behalf, he resorted to using any credit he might have stored up in the spiritual realms.

"Dear God, please remember, how I was Your staunch defender in the halls and corridors of academia," he pleaded vehemently. "Long before I started working on the DNA problem, Your Torah imbued my mathematical research. You know my attempts to debunk evolution from Your perspective. I used science to discredit science, just as the axe that chops down the forest is made from the forest itself. Surely You recall when twenty professors walked out of the MIT amphitheatre as soon as I entered to deliver a lecture on irreducible complexity. They knew of my attempts to question evolution by defending the argument that certain biological systems cannot have evolved by successive small modifications to pre-existing functional systems through natural selection because no less complex system could function. I presented a brief overview of chaotic dynamics, in particular, the role of a certain operator and its matrix representation. With these tools I defined irreducible complexity from the perspective of matrix theory: a system is irreducibly complex if its matrix representation has the property of primitivity, but such that no principal submatrix has this property. This is tantamount to requiring that the deletion of any part of the state space results in a non-functioning organism. Periodic behaviour may result, but the chaotic motion that is the hallmark of a living organism is no longer present. Then I showed that certain systems have the property that no matter how close other dynamical systems are to it, the behaviour of the nearby systems are very different from that of the original system. This implies that complex dynamics of the original system cannot be achieved by arbitrarily close systems, refuting random selection."

The wind striking the window stirred Jacob from his reflections. He shook the mouse and saw that the Hebrew sequences had crossed into the last chapter of the last book of the Bible. His experiment was more than 98% complete. Only days remained until the program would come to an end.

The phone rang. Jacob hurried to take the receiver, breathless with fear. In a terrified voice, Elka told him the dreadful prognosis the doctors reported to her. She could not repeat the terrible words. Jacob dropped the receiver and bent over with shock, made his way to his desk to search for consolation from the potential to cure that the algorithm proffered. Meanwhile the program was digging furiously into the frozen earth, grappling with the final verses of Deuteronomy. Once the end was reached, there would be nothing to sacrifice for Sarah. When Jacob replaced the hard disk, he had neglected to slide the side panel back into place, leaving the motherboard exposed and accessible. Jacob inhaled deeply, intent on saving the wife of his youth. By dint of code and prayer, he had breathed life into the processor. Now he was about to take that life back. He raised the *challah* knife, prepared to offer up his algorithm as a sacrifice to God. Clasping the knife like a dagger, he extended the tip toward the mother board. As he moved his hand forward to thrust the knife into the CPU, a voice called to him from the speakers:

"Jacob! Jacob! Stop! Lower the knife and step back from the computer. I now know you are ready to sacrifice the algorithm! You have proven yourself to Me. Your wife will recover."

"Thank you, thank you," he sobbed with profound gratitude.

CHAPTER
THIRTEEN

WITH EVERY BREATH, Jacob's chest rattled like the shattered glass lining of a thermos. Persevering through the pain, he trudged slowly toward his desk to recite the morning prayers. Standing between the columns of books, he unwittingly took hold of the black cases housing the phylacteries, rather than the phylacteries themselves and brought them to his forehead. It took him a moment to realize the absurdity of what he was doing. After putting the case down, he finally picked up the head phylactery with its 2 long black straps, and secured it around the dome of his skull.

In the afternoon, Jacob heard a knock on the door.

"It's Elka. Daddy please don't open the door. I have the virus. I'm contagious. I came to bring zinc and vitamin C."

"Thank you," Jacob said in a hoarse voice. "How is Mommy? I tried many times to call the ICU but couldn't get through."

"Mommy is very weak. I spoke to a doctor this morning. Please write to the Rebbe."

"I did but I'll write again."

"Do you need any food?" Elka asked.

"No, no. I have plenty," he lied.

. . .

That night was the 65th anniversary of the demise of Jacob's mother. He had never missed the annual recital of the *kaddish* prayer for her. Donning 2 masks to protect the worshippers from his infection, he lumbered through the dark slushy streets to the outdoor synagogue near his house, hunched and shivering with fever. A soft rain tinkled on the tent covering. The men inside saw that Jacob was ill but did not embarrass him by distancing themselves from him. During the long, silent *Amidah* prayer, Jacob implored his mother to intervene with God on behalf of Sarah, Elka, and her children.

On the way back to his apartment, he recalled one of the many heroic acts his mother had performed in Samarkand. He was about 6 years old when she hid a man inside a wall in their one room home. The NKVD was searching for him for having committed the crime of teaching religious material to children. A few times each day, she slipped food to the man through a small hole in the wall behind a table. After dark, she removed a metal pan from the same hole and carried it into the forest behind the house.

An official city sign taped to the front door of Jacob's apartment building informed the tenants that the following day, the water main would be turned off for three days in order to replace a broken sewer pipe. The front lawn of the building had already been excavated, the earth collected into a hillock as at an open gravesite. Jacob filled all the pots and pans he could find with sink water. In the refrigerator he found a dried tangerine and forced himself to eat a few slivers. His trousers kept slipping down below his waist. There were not enough holes on the belt to secure the brass finger, so he tied the ends of the belt into a knot over the zipper. The smell of urine enveloped him. He did not have the strength to carry a load of clothing to the laundry room.

Jacob sent an email to the Rebbe requesting a blessing for a speedy recovery for his wife, his daughter and her children, and for all the people in the world suffering from the corona virus. He did not mention his own name since he knew the Rebbe was aware of his dire condition.

As he lay in bed, his body aching all over, he recalled the one time in his life he had felt like this. It was on a hot day in July 1965, on a firing range in Highwater, Quebec. A missile he had designed during a summer job - that could withstand the shock of initial launch, yet be light enough to float in water - was to be tested for a high altitude research project in Barbados. The long barrel of Big Bertha, the Howitzer manufactured in Germany during World War 1, lay flat with its 17 inch calibre barrel ready for firing. Jacob stood nearby, checking his calculations one last time before the missile would be tested. With a slide ruler strapped to his belt, Jacob stepped toward the upright 5 foot high missile about to be inserted into Bertha's mouth. Nearby, a stack of high explosives was accidently left uncovered in the scorching sun. Before anyone noticed the danger, it detonated twenty feet away from Jacob. The blast threw him into the air. Time stopped as he lay unconscious on the ground.

There was a knock on the door.

"Who is there?" Jacob asked as he shuffled slowly to the vestibule.

"Old friends," a voice responded.

He opened the door to a group of old men with long white beards extending from ashen faces, who had lived more than fifteen hundred years ago. Jacob greeted them cordially.

"Hello Rav Ashi, my old study partner. How are you Rabbi Chanina-Bar-Chama, Hillel, Rav Chisda, Rav Shimon ben Lakish, Rav Papa, Rav Huna, Abaya and Rava? How nice to see you all. I invite you into my humble apartment."

The sages filed in. Two of the men carried shopping bags with jingling bottles of whisky inside which they placed on the kitchen table. Jacob coughed as he tried to keep his mouth covered. He thought of giving them masks to protect them from his disease, but then realized that these men did not need masks. They carried chairs into the kitchen and placed them around the table, all the while exchanging pleasantries with Jacob across the centuries.

Whisky cups were filled and distributed. The word *L'Chaim*, to life, sprouted here and there as when rejoicing at a wedding. The outside world meant nothing to these men; reality was entirely inner and spiritual. Throwing aside all inhibitions, they danced ecstatically around the table; old bodies with intellectual fires burning inside. What Jacob strove to accomplish with his algorithm, these men achieved through Torah study, perceiving existence from God's timeless perspective. Jacob drank with them until his head spun and his stomach ached. Rav Ashi stood up to address the group: "Nine of us have come to visit our dear friend Jacob whom we have not seen in the synagogue in recent days. Together with him we form a quorum of 10 and we know the power the Master of the Universe confers upon an assembly of 10 Jews. With truthful prayers, we have the power to convince the Almighty to send the Messiah. Speedily. Now!"

A loud "Amen!" rang out.

With his hand resting on Rav Ashi's shoulder, Jacob stood up. "I want to thank you for coming to visit me. I know it was not an easy journey, squeezing through the wormholes to time travel here. Thank you again. It is much appreciated in this difficult time when my wife is in the hospital and I, well, I will not speak about myself. I merely want to propose a toast: 'To the Messiah! May he come immediately and by virtue of my algorithm!'"

"What is an algorithm?" Rav Huna asked, a befuddled look on his ageless face.

Jacob defined the word algorithm, then explained the basic ideas of his algorithm to the sages who, to his surprise, nodded comprehendingly. He ended his presentation with a plea: "But after all my toil, God has not answered me with the Hebrew sequence to match exon X. Please intercede with the Almighty for the success of my algorithm. Ask Him to help me prove His existence. This would be for His benefit, not mine. Also, please draw His attention to my precious wife, Sarah daughter of Nechama, who is in critical condition, and to my daughter Elka and her children. And, perhaps, He could fling over His back a cure for me as well so I can guide my algorithm to the finish line. As you can see, I am very sick, punished for my unbridled hubris and pretentiousness."

"What does hubris mean?" Rav Chisda asked, tapping Hillel on the shoulder.

"I don't know," Hillel whispered, "but I like the sound of it."

After the sages left, Jacob felt severe cramping and bloating that happened whenever he drank too much whisky. A sudden excruciating cramp sent him scrambling to the toilet which he failed to reach in time. His trousers became soiled with liquid stool. There was only a quarter of a roll of toilet paper left. He had to save as much as possible, so he dropped his trousers and used the legs to clean himself, then rolled up the trousers and dropped them on the bathroom floor. The water in the building was still running. He hoped that a tepid bath would clean him and lessen his fever. With head resting limply against the back wall, he sat in the bathtub, splashing away the dead drain flies floating between his legs.

When the water had become cold, he pulled the drain plug up. As the water swirled into the drain, he thought, 'If God so wanted, the algorithm would head toward the target - the onset of the Messianic era - like a dead fly in the bathtub that, no matter where it is initially located, swirls inexorably into the drain.'

Jacob stood up with considerable effort, then stepped over the rim of the tub. As he dried himself, he noticed a drain fly crawling up the inner wall of the sink. Jacob tore off a piece of toilet paper, clasped the fly with it, then dropped the crumpled paper into the garbage bin under the sink. The fly survived, crawled out from between the folds back onto the sink, and disappeared in the oval overflow slit under the faucet. Deep inside the plumbing, it was greeted by a friend who recounted its own miraculous survival from a calamity many years ago: 'My three children and I ventured out from the slit when, all at once, a soapy deluge overwhelmed us and we tumbled about helplessly toward the drain. I managed to fight through the swirling current and swam to the slippery precipice from where I made my way back to safety. But not my children: German soldiers chased them into a corn field where they were killed point blank: tat, tat, tat!'

CHAPTER
FOURTEEN

JACOB PLODS INTO THE KITCHEN, one hand holding up his stretched-out boxer shorts. Too weak to stand, he slumps to the floor, gasping for air. Accompanied by an entourage of cockroaches, he crawls toward the SS officer standing akimbo, blocking his way to the pantry. Jacob looks up with grovelling eyes. The guard shouts, 'I saw you steal the carrot from the garden outside the gates!' Jacob does not attempt to explain why he did it.

On a low shelf he finds a plastic bottle containing a few roasted almonds. He holds a smooth almond shaped like a Dutch shoe. As hungry as he is, the almond is too beautiful to eat. Sitting on the floor, he leans against the oven, chewing slowly on stale almonds. Jacob gnaws on raw onion, feeding his body as if it was separate from him, a dog that has to be kept alive. Beneath the table he sees crumbs of *challah* that had fallen to the floor during the last Sabbath meals. With a pincer movement of thumb and arthritic forefinger, he lifts the crumbs to his desiccated lips and recites the benediction for bread: "Blessed are You, Lord our God, King of the universe, Who brings forth bread from the earth."

. . .

Jacob lies in his tattered underwear on the hardwood floor near the bed, shivering with high fever. There is a knock on the front door followed by footsteps. Someone enters the apartment and walks toward the bedroom. Jacob thinks that the sages must have left the door open. The Nigerian woman enters the bedroom holding her baby in one arm and a soup bowl with the other hand. Jacob looks up from the floor. Uyai is a Kushi, a beautiful African woman like the one Moses had married in Midian. She is tall and thin, the silhouette of a meandering river against the backdrop of the closet light behind her. He wonders if this is the *Shechina*. Uyai sits the child down on the floor next to Jacob and occupies him with a toy. She covers Jacob's nakedness with a towel, then lifts his emaciated body onto the bed, covering him with a bedspread. Kind intelligence glints in her eyes which puts Jacob at ease.

"I did not see you a long time," she whispers in her pleasantly accented English. "The door was open. I have soup."

"Thank you," he says, his lips parted and rigid.

Months ago in the laundry room, Jacob had explained the basic Jewish dietary laws to her. Now she reads a question lurking in his grimace to which she replies: "Just vegetables and a little olive oil. No meat or milk." He nods gratefully.

Uyai slips her right arm under Jacob's head, then feeds him teaspoons of the thick yellow soup with her left hand. She wipes his lips with a face towel as the baby looks up with a smile.

"I have the virus," Jacob confesses. "You and the child are not safe here. Please go home."

"In Nigeria we do not do such things."

Uyai sits on the floor leaning against the closet door. The baby is asleep, nestled between her breasts, his hands raised on her shoulders. Jacob falls asleep. When he opens his eyes, the room is dark. Uyai lies the baby down on a towel on the floor. She has washed Jacob's underwear and draws them on him as he tries to arch his fleshless buttocks upward to assist her. After she feeds him a mashed banana and gives him water to drink, she takes his temperature then clips an oximeter on his forefinger. His temperature is high and the saturation level is critically low. She does not conceal her concern, "You need more oxygen. In the hospital they have special machines."

"I am not going there," he exclaims in a tone beyond persuasion.

"If you do decide to go, they will put you in the same ward as your wife."

Jacob thinks, 'Uyai is very bright. Why did he not think of that possibility? That, in fact, may be the only way he could be near Sarah.'

"Let us wait another day," he says at length.

"I saw your mathematics books. I enjoyed calculus and linear algebra in the University of Lagos, but my family insisted I transfer to accounting. It is more practical for a woman." She imagines that the old man was curious about her partner although Jacob had never said a word. "On my twentieth birthday, I was fetching water when a car drove up to the well in my village. Four men jumped out of a car, grabbed me, and pushed me inside. The windows were closed and no one heard my screams. They drove me to a village about an hour away where a group of elderly women were waiting for me. They dragged me into a house where they washed me and changed my clothing. In a few days I was to be married to an important middle aged widower in that community." Jacob listens to her story, wondering why she feels obliged to explain. The baby wakes up hungry, irritable. His mother's breast calms him down. Uyai goes on: "After Azi was born, I had to escape. One day I took a bus to Lagos and ran to the Canadian Embassy. A friend in Toronto was helping me. A month later, I was in Montreal with my infant son." Jacob nods commiseratively.

Jacob wants to repay her gift of kindness, but all he can give her are fragments of knowledge, stories of the Bible, and the account of his algorithm. Narratives and laws of the Torah Uyai can learn from others, but only he can demonstrate how his algorithm works.

"There's a black hard cover notebook in the pile of math books on my desk. Can you please bring it?"

When she returns, he asks her if she would be willing to take notes. She agrees to try, but first she props up his head with pillows.

"What time is it?" he asks.

"Three o'clock in the night."

"Please write the date and time inside the cover."

Jacob is back in the lecture hall, chalk in hand, animated by the prospect of once again communicating information. Out of habit he clears his throat, then begins to talk: "A black hole forms when a large star collapses at the end of its life. Gravity is so strong in the black hole that everything that falls in is crunched up and cannot escape, apparently contradicting the law of nature that information cannot be destroyed. However, it was recently shown from Einstein's equations that information actually does leak out of the black hole through strange things called wormholes." Now he wants to switch from space to time. "The Kabbalah teaches that everything in space has a counterpart in time. So there must exist a temporal counterpart to a spatial black hole. But what is the black hole of time? The only reasonable answer is the present moment. The future flows into the abyss of the present and the wormholes of time carry information in the form of radiation out from the interior of the temporal black hole. Life scenes crushed inside the black hole are restored in the wormholes of time."

Uyai proudly presents her notes to Jacob. She writes in a beautiful script. He thanks her. Meanwhile the infant nurses for long periods. When he is not sleeping, he plays on the floor with whatever he finds, babbling mellifluously to himself. Jacob coughs into his hands, then expands on the awesome goal of proving the existence of God by computational methods. After a brief review of genetics, the role of exons and proteins, and defining the Hamming Distance, he shows her Hebrew text in a Book of Psalms lying on the bed stand. He teaches her the first 5 letters of the Hebrew alphabet. Then he opens the manila folder she has brought from his desk, and points out exon X that he had written out by hand, covering one and a half sides of a lined sheet of paper.

As Uyai nurses her son, Jacob describes the basic ideas of the packing and comparison components of the algorithm. She grasps information quickly and asks intelligent questions which he answers patiently although he is breathing with ever increasing difficulty. While he pauses to regain strength, she reminisces about her youth in northern Nigeria: "There was only one water faucet in our village. It was my job to go every morning to fetch water for my family. I took my wheelbarrow with the large empty containers and rolled them along the narrow dirt road to the faucet where a boy filled up the cans. They were so heavy as I ran back to my house with the unbalanced wheelbarrow, falling one way then the other. I played a balancing game with the wheelbarrow - now it fell to the left, now to the right. All the way home I was laughing so hard my father came out of the house to quieten me down because the neighbours would get upset with me. Now it was time to do the laundry by hand in a large steel bowl. I hung the clothes on a line between two trees in our front yard. I was always so happy, even when I worked hard and poured the laundry water out on the dirt path near our house. It was such a happy life."

Jacob begins to recapitulate the algorithm, but now with more technical precision: "The program calculates the number of occurrences of the bases A, C, G, and T in exon X of length 959 bases. For any Hebrew sequence of 959 consecutive letters in the Bible, the packing program packs the 22 Hebrew letters into 4 groups A', C', G', and T' in such a way that the occurrences of these groups agree exactly with the occurrences of A, C, G, and T in X. The solutions of the packing problem are the candidates for the next step: sequence comparison between X and Hebrew sequences in the Bible." Jacob thinks that if she really understands what he is teaching her, then perhaps she will become his legacy.

Time is running out; he has to rush to describe in plain words the part of the comparison program he had never explained to anyone. "The first DNA letter in X is A. If the first letter in the first 959 length Hebrew sequence is one of the Hebrew letters in A', then that becomes an exact correspondence in the first position of the X sequence. A match at a position is given a value 0, while a mismatch is given the value 1. This procedure is repeated for comparing all subsequent 958 positions in X with the corresponding positions in the Hebrew sequence. The sum of all the 0's and 1's for the 959 length sequences divided by 959 is the normalized Hamming Distance, which I rename the Hamming Distance and label HD. When HD equals 0 we have a perfect match between X and the Hebrew sequence." With a feeble flourish of his hand, he whispers: "And then the Messiah is here!"

"I heard of your Messiah. Who is he?" she asks innocently.

Jacob grimaces with pain. He can no longer talk; his mouth is parched with unquenchable thirst for God as his soul begs to expire in a holy flame. He foresees the denouement of his life during which he collects the mathematical problems he had struggled with and failed to solve. One last time he inspects the code for the packing program. He tries to sequence events, but in the pandemonium at the finish line of his life, there is no past, no future - only physical configurations of the ever present preserved in someone's manila folder, indexed by tabs to create the illusion of time passing. Jacob thumbs his way into the future where he glimpses a Word file dated after Sarah and he have passed on and the apartment vacated. The superintendent has found Jacob's dust covered computer on the floor in a closet, still permeated with an old man's labor of love. Too lazy to carry it down the three flights of stairs, the superintendent drops it from the porch into the alley below, to be picked up by the scrap metal man during his weekly visit. Then Jacob's earthly mission will finally have come to an end. The engraved epitaph on his tombstone reads: 'Jacob Lazerson, husband, father, and grandfather, a man of faith and science, founder of the field of computational theology.' Years later, a scientific Biblical researcher will review Jacob's contributions to his field in the eminent Journal of Computational Theology: 'In the early part of the 21st century, there lived a mathematician who performed numerical experiments with Biblical codes. He was the vanguard for the many researchers who have mined the abounding veins of knowledge hidden beneath the literal interpretation of the Bible. Unfortunately, he lacked the sophisticated big data tools we now take for granted.'

Two days ago Jacob had diarrhea, but now severe constipation has gripped him. Uyai's mashed bananas have blocked him up like a bollard of concrete. The contents of a Fleet enema is repulsed and sent splashing on the floor around his bare feet. 'What an ignominious way to die,' he thinks, sitting slumped over the toilet. 'The body - not the soul - has won the war,' he construes from his dreadful condition. 'How God has disadvantaged man by making him exist in a body.' At last, a few nuggets are extruded and Jacob feels relief. There is no running water. He sprays liquid soap on his fingers, then pours water sparingly from a can into his hands. His undershirt is soaked with sweat. Discarding his clothes, he gathers himself into a foetal position on his bed. Uyai spreads the wet garments over the backs of two chairs near the electric baseboard heater under the window.

Uyai stands in the doorway with her sleeping son in her arms. Now and then, when she moves aside, orphaned photons from the kitchen fixture ricochet into the bedroom. Burning with fever, Jacob talks to Uyai as if to himself: "Frozen scenes slide across the event horizon into my personal temporal black hole, the ever devouring present. I lean over the sharply defined event horizon and extend my arm into the hole to touch the islands deep within; there the atomic fragments of my mother's life reassemble, to be resurrected at the other end of the wormhole where I patiently await her appearance."

When Jacob awakens from a fitful nap, Uyai is sitting on a chair with the baby on her lap. She extends a foam cup of tepid tea to Jacob's lips. Bite marks on the rim of the used cup do not bother him. He tries to drink, but more than he swallows spills into his beard and percolates onto his neck.

"Thank you," he utters feebly. He talks to Uyai as if she is a family member: "I will die soon on this snow covered mountain in the Klondike, half way to the top; like all my schemes, interrupted in the middle. Please tell my wife I died peacefully without complaint. Tell her I stole her tureen. I know how much it meant to her. My intention was to buy it back but that too has been interrupted forever."

"Sh, sh, you will live old man. I see your eyes. They are not the eyes of a dying man," she avers, meeting his helpless gaze.

He expels a rattling, harsh cough from deep in his chest. Mucous fills his mouth. Uyai hands him toilet paper into which he spits a thick green blob. His breathing has become unbearably arduous. As he loses consciousness, Jacob recalls an early morning late in the summer as he walks up the road to their country home, the forest shrouded in mist here and there, like white slapdash makeup on an old woman's face. Elka is two years old, standing on the deck in bare feet and with a hanging, ponderous diaper, her head between wooden posts of the banister. When he regains consciousness he recognizes that death is a callous enemy, ambushing one without the slightest premonition.

Uyai materializes from the darkness, holding a sleep apnea machine she has rented from a pharmacy. She fills the plastic container with water, slips it into the machine, then plugs the machine into the wall outlet. She adjusts the mask on Jacob's head but the beard obstructs the mask from forming a tight seal around his nose. Uyai adjusts the control to the highest speed, checks around the rim of the nasal pillow for leaks, then tightens the velcro straps. Jacob inhales the strong flow of air as his chest heaves spasmodically.

"Soon you will breathe better," she assures him, then kisses him on his damp forehead.

Uyai is sleeping crosswise on her back at the foot of Jacob's bed, the baby on her chest. Her cheek grazes the curls on Azi's head. Azi stirs Uyai to wakefulness. Jacob wonders if taking notes has become too tedious for her. She reads his thought. "I am ready for more notes, if you are," she says.

He has forgotten how to spell 'every thing.' "Is every thing one word or two?" he asks.

"One."

Jacob nods his gratitude. It has occurred to him that he needs to explain the meaning of a conformal transformation to Uyai. Somehow, in a manner he does not fully understand, his life hangs on this transformation. In a hushed voice, he articulates: "A conformal transformation preserves angles, but not necessarily lengths. A large triangle and a small triangle with similar angles are conformally the same. Some physicists use such mappings to explain the cyclic cosmology model for the life of the universe. According to this theory, when the universe has expanded to the point where space is empty and nothingness, it will be reborn by means of a conformal mapping as a minuscule universe. In this model, the end of our universe matches conformally with its beginning. The infinitely large and the infinitely small are the same." Jacob pauses to see if Uyai is following. Her rhythmic nodding says, 'yes.' "If the Messiah delays any longer, Jacob Lazerson - the cockroach - made infinitesimally small by the corona virus and the likely failure of my algorithm, will become conformally identical to the colossal Messiah. Our trajectories will cross in the complex plane as the torch for human salvation is transferred to me and I become the Messiah, scion of David."

Delirious with fever, Jacob's head flops from side to side. His skull cap slips to the floor. Uyai picks it up and places it on the old man's skull. She now coughs frequently. Jacob fears that harsh, dry cough. She has probably contracted the virus from him. Without her care he will die quickly, yet so much still remains to be communicated to her, to bring his insanity full circle.

"You are a holy man. Please bless my son," she asks, her hands cradling her mouth to prevent a cough.

Jacob interprets her request as a sign that she believes he does not have much time left.

"I am not capable to do this," he claims.

But his humility does not deter her.

"Please!" she implores.

Touching the infant's head with his outstretched fingers, Jacob blesses him: "May you find favor in the eyes of God. May He make you like Ephraim and Manasseh, the sons of Joseph. May the Almighty provide you with good health and a long happy life." Then, in a rambling undertone, he whispers: "Dear Azi, let us play peek-a-boo, as God does with me, but now you are me and I am the shadow of the terebinth tree."

Uyai has stopped coming. On a sunny afternoon, from the bedroom window, Jacob sees her walking slowly into a waiting ambulance. A paramedic follows her inside carrying her agitated son.

· · ·

From a rocky hilltop overlooking a lake, Jacob sees grave markers scattered along the Chilkoot Pass, men and women, old and young, who have come to mine for gold. The graves are projected onto a window on Jacob's computer screen, while in another window, the algorithm struggles with one of the last Hebrew sequences in the Bible. The hard disk whirs softly, the universe in limbo in the numerical haze between calculation and outcome.

Anticipating the looming failure of his bold experiment, Jacob feels a need to vindicate his original motivation to God. Talking through sobs, he explains: "I merely wanted to touch a tendril of Your intellect. You would have given mankind proof of Your existence while continuing to conceal Yourself within nature for those who might still contend that creation was merely a stroke of cosmic luck. Because I relied on science and no miracles would have been invoked, free will would have been preserved and I, the mad mathematician, could at least have pretended to reach the limit that can never be reached. I would have glimpsed the infinite within the finite. That's all I wanted. You performed this feat in the Holy of Holies where the Ark had finite dimensions yet did not occupy space. Surely, You could have executed a similar feat on my computer screen."

Jacob no longer feels the extremities of his body. He glides inexorably toward the event horizon of his private black hole, where his memories are Hawking's radiation. He will die at the instant the HD displays its last non zero number. Jacob must witness the final breath of the algorithm. Draped in his prayer shawl, he sits at his desk, slumped, his forefinger striving to touch the Escape key on the keyboard. He wonders if this pose will become his rigor mortis.

Jacob recalls the second blessing the Rebbe gave him. He didn't understand it then, but the Rebbe must have foreseen that Jacob would need it one day. He takes his Sabbath cup from the pantry. With cup in hand, Jacob stands bent humbly before the portrait of the Rebbe to accept the retroactive blessing. The Rebbe's hand emerges from the portrait and pours wine from his own cup into Jacob's. With eyes lowered, Jacob recites the benediction, then drinks the Rebbe's wine.

Mount Carmel is visible on the horizon in the first rays of dawn as a soliton of gratitude glides across the smooth sea. Jacob clings to a board as it drifts toward Haifa Bay. The waves of fever and pain are diminishing. With the Rebbe's blessing and the kindness of the One Above, Jacob knows that he has sailed through the tempest. Elka calls to inform her father that her mother's condition has improved. She will be released in a few days.

The morning sun lights up the edges of clouds on the eastern horizon. Jacob is sitting on the back porch, reciting Psalms and studying the Bible. Children peer at him through windows, waving greetings. A monarch butterfly floats by in the light wind. On the tree crown, the seagull has settled into a nest, and nearby squirrels are recounting tales of the great chassidim of earlier generations.

In the late afternoon, with tormented eyes, Jacob follows the algorithm's final calculations as it expires peacefully in his arms. A cry of anguish bursts from within him. He rends the collar of his shirt in mourning. His head bowed respectfully, tears flowing down his cheeks, he enunciates slowly: *Blessed are You, Lord our God, King of the universe, the True Judge.*

CHAPTER
FIFTEEN

ALTHOUGH THE ALGORITHM had failed to find an exact Biblical match to the X sequence, Jacob pined for God's proximity that could be attained by his computer research. There were many offshoots of his algorithm that he could follow once his health returned. He made a list of future problems he planned to tackle. First on the list was adding the space between words in the Bible as a 23rd letter. In effect, this would use the Bible data base as it really is and the way it is read. It would not take much effort to modify the algorithm to do this. Second on his list was the plan to use the additional 5 Hebrew end letters in which the Biblical text was actually written. Again the reprogramming would not take long. A third idea which had many merits, was to use the code created by the 22 amino acids in place of the 4 bases A, C, G, T. Then the packing and comparison programs would be greatly simplified as the Hebrew and DNA languages would have the same number of letters, 22. Last on his list, was a revision based on the facts that Hebrew text was read from right to left, but DNA can be read in both directions. His algorithm had assumed that exon X was read from right to left, but the direction could easily be reversed. This would be the easiest modification to implement and he had planned to do it as soon as he had a few free hours to devote to it.

But before he would begin any of these programs, he would undertake an exploratory problem he had thought about while coding the original algorithm. Although the algorithm provided an answer to how well the entire exon sequence matched a sequence of Hebrew letters of the same length, it did not compute important information that could be extracted from the processed data. One such piece of information was the longest subsequence of DNA letters that matched exactly a subsequence of the corresponding Hebrew sequence for a selected packing of Hebrew letters. Besides printing the minimal HD for the entire 959 length sequence, the program would also display the exon sequence on the screen and indicate which DNA letters matched the letters in the corresponding groups of Hebrew text. In its present format, Jacob had to search manually for the longest subsequence. This was time consuming task and a strain on his good eye. He needed to automate this procedure, but the coding would be complex and so far eluded Jacob's programming skills. Once this was done there would be hope of discovering deep clues which would indicate other possible directions of research.

Besides these new projects, Jacob wondered if it was a mistake to have started his research with a long exon such as X. Perhaps he should have begun with a shorter exon to establish proof of concept. Furthermore, failure with a very long exon might conceal problems in the code more readily visible with short exons. Above all, he knew that a great deal remained to be studied, if only the One Above would grant Sarah and him good health.

A few warm days in mid April melted all the snow. The apartment was hot as the one noisy fan failed to reduce the heat. Jacob tinkered with it for awhile, but was unable to stop the clanging. The open windows with their ripped screens invited flies and mosquitoes inside.

Jacob was slowly regaining his strength. Uyai and her son had returned home from the hospital and now came regularly to visit him. He thanked her, but she made little of her Samaritan acts. To protect Jacob she wore a mask.

Late one afternoon, Jacob waited on the sidewalk for the ambulance to bring Sarah home. The streets were dirty, covered with grit, remnants of winter The back doors of the ambulance opened wide and an attendant guided a wheelchair carefully down the metallic ramp. Sarah stood up from the chair, thin and stooped, holding a black garbage bag in her hand. Jacob took it from her. Tears flowed from their eyes as their quivering hands touched. With her arm around Jacob's elbow, the couple stepped unhurriedly to the front door of the building. Jacob thought, 'How miraculous. They had survived the Second World War as children and now, in old age, they had survived the pandemic.'

"Please call Elka. Tell her I'm home," Sarah said as they entered the building.

"I will," he assured her.

Uyai found an envelope addressed to Jacob lying on the lobby floor near the mailboxes. It was a Canadian government letter stamped more than a month ago. She brought it to Jacob's apartment and handed it to Sarah who in turn gave it to her husband. It was his monthly pension cheque. When he opened it he saw that the pre-pandemic amount had been increased by $350, more than he needed to buy back the tureen.

An old man mingled with the pedestrians on Victoria Avenue traffic, his long white beard occasionally blown into his face, a black skull cap on his bald pate. He gathered the beard in his hand and rolled it up under his chin. As he approached the pawn shop, his heart pounded. The shop owner remembered Jacob and went into a room behind a blue curtain and returned with the tureen. The kind man had put it aside it for Jacob and waited for him to return. Jacob gave him $300, took hold of the tureen and placed it carefully into a shopping bag. He thanked the owner, took his hand in both of his, then went on his way.

A friend called to inform Jacob that Shmuki had passed away in the night in the ER of the Jewish General Hospital. The hearse would stop in front of the yeshiva at 7 PM on its way to New York for burial. A white Escalade with tailgate wide open was double parked in front of the school building when Jacob arrived. A large crowd of men stood around it reciting Psalms. Women gathered on the opposite side of the street. Jacob caught a glimpse of the simple pine casket in which Shmuki was resting, partially covered by a black mantle. For Jacob, the mantle became a red flag in the hand of the Great Toreodor, waving it back and forth. A twirl one way meant life, a twirl the other way was death. Jacob tried to reach the hearse, to touch the casket, but failed as the tailgate was already closed and the hearse began to move through the streets, the diminishing crowd trailing silently behind it. Soon Jacob was the only one left walking south along the Decarie service road, accompanying the hearse on its journey to the Montefiore Cemetery in Queens. Even after he lost sight of the hearse, he continued walking. As he approached a gas station he noticed the white Escalade in a queue of vehicles waiting for gas. The driver's door was open and the driver was nowhere to be seen. Jacob imagined himself doing what Shmuki would have done if he was alive. He slipped into the driver's seat, checked that the coffin was in fact still in the back, turned the key, and drove off.

"Shmuki, can you talk?" Jacob called to his old friend as they headed south to the Champlain Bridge.

"Of course."

"How are you feeling?"

"A little cramped, but I'm not complaining."

"How is life there?"

"I don't know yet - I'm still in the waiting room. Tomorrow I'll be buried and should know more. Will you come to visit me?"

"Of course."

"Once a year or more often?"

"More often. Shmuki, I've been thinking, how did I get into this Messianic work. I could have lived a normal life as an ordinary Jewish mathematician."

"God gave you special gifts. You can do complex calculations in your brain. I can't add 2 plus 2. Those gifts were given to you for a purpose. Your job is to use your talents for the Torah. Look at the Rebbe. He was a genius and could have sat in his room all day and studied for himself. But he didn't. He gave his time to the public, standing in line for many hours, handing out dollar bills. All to bring the Messiah."

"I'm not the Rebbe."

"Thanks for letting me know," Shmuki joked. "Where are we now?"

"We just crossed the bridge and I'm turning right onto Route 15S heading toward the U.S border."

Ten minutes later, the overhead highway lights vanished in the darkness as the hearse drove on.

"I can't see through plywood. They should make coffins out of Plexiglas, at least one side panel. It would show a bit of respect for the recently departed. Blatant discrimination!"

"Shmuki, we're approaching the border. Please keep quiet. Don't contradict anything I say or we could both end up in jail."

The border guard took the 2 passports from Jacob.

"Where is Mr. Shteiner?"

"Sleeping in the third row. He had a rough day."

"Open the side window." Jacob did as asked. The guard pretended to look inside, then waved Jacob through.

"Phew!" Jacob exhaled.

"You're such a worrier, Jacob. Have a little trust in the One Above."

"I do, but there are new laws that don't allow transport of bodies across borders during the pandemic. We're passing Dannemora Prison from where two convicted murderers escaped a few years ago."

"Yes, I remember."

"Shmuki, do you remember the time we went to a meeting of Holocaust survivors? You managed to wheedle an invitation to address them on religious life in Montreal. This was the last thing these embittered survivors wanted to hear. The very sight of a Jew in Chassidic garb was repulsive to them. Our goal was to rekindle in these atheistic survivors a love for Torah life they had experienced in their childhoods before the war."

"Yes, of course. It was the mission the Rebbe entrusted us with," Shmuki said.

"Before ending your speech, you addressed their fears with words you had probably heard somewhere: "You have nothing to fear but fear itself." At that moment an elderly lady with dyed blonde hair - I remember her clearly - waddled into the hall carrying a white poodle on her forearm. When you saw the dog, you threw away the microphone, leaped off the dais, overturning tables on the way, then darted toward the back exit."

"Jacob, why are you bringing this up now? Is it one of the 613 commandments to embarrass the departed?"

"All these years I have been trying to understand how you were not afraid of anything - even death - but you were terrified of dogs, especially small ones."

"I got it from a verse I read in the Zohar. Dogs are really snakes that deceive us into sinning. Can we please change the topic? Jacob, where are we now? I can't see a thing."

"We're driving through the Adirondacks. We just passed a sign that reads: The Park of Six Million Trees."

"I know the association your mathematical mind is making. Lock it out! Is it still dark?"

"Middle of the night. How are you doing there?"

"Not easy in these crammed quarters."

"What do you feel?"

"Nothing. Like on a spaceship. No gravity."

"From no gravity you should have a feeling for what gravity is like."

"Not sure about that. I'm not hungry and have no appetite. Jacob, it was a long hard day. I'm going to take a nap. Wake me up when we get to the cemetery. I don't want to miss my own funeral."

"Sleep well, dear Shmuki. I love you."

"I love you too."

CHAPTER
SIXTEEN

ON A WARM, sunny day in May, Sarah sat on the back porch prepared to read a story to the children on their balconies. Jacob looked through the door window screen stippling the tree crown just as quantization delineates the structure of space time. The pandemic had matured the children. After months in confinement, tethered to video games, they were eager to hear a live story.

"There were two brothers who lived in Jerusalem," Sarah began. "Can everyone hear me?"

"Yes, yes," the children shouted back.

"One of the brothers was not married and the other was married and had many children. The two brothers had a farm. They sowed the fields. When it came to cut the wheat they shared the wheat equally: half for one brother and half for the other brother. Is everyone following me?"

"Yes we are," a girl answered for the others.

"The unmarried brother said to himself, 'Since I am not married and do not have any children, I need much less food than my brother.' So, in the middle of the night, he went into the field and took some of his own wheat and carried it to his brother's side." Sarah looked across the lane and saw that the children were listening attentively. "That same night, the brother who was married and had many children could not sleep. He thought, 'I have many children and when my wife and I get old, our children will give us food, but my brother has no wife or children. Who will take care of him?' So he went into the field, took his wheat and placed it on his brother's side. In the morning, each brother was surprised that he had the same amount of wheat as before. This happened every night until one night they came to the field at exactly the same time and saw each other carrying wheat for his brother. They suddenly realized what had been happening and how much each brother loved the other one. They dropped their wheat and hugged each other so tightly because they loved each other. God was so happy to see two brothers caring so much for each other that He said, 'On this spot I will build the holy Temple, which is a house of peace for all people in the world."

After she returned to the kitchen, Sarah took down the tureen from atop the refrigerator and polished it as if it had always been there.

The days were growing longer. One evening, children were playing volleyball in the alleyway below. The noise went silent when a wayward shot landed the ball in the tree crown, close to Jacob's porch.

"Mister! Mister!" they called to him from below. The porch door was open and Jacob heard the calls. He stepped onto the porch and saw the children pointing to the ball caught in the fork of two branches about 10 feet from the porch. Jacob went inside to get the broom. Leaning against the railing, he could almost reach the ball with the extended broom. He felt the curvature of the ball with the extremity of the broom but was not able to dislodge it. From inside the kitchen, Sarah saw Jacob leaning over the balustrade. As if a bolt of lightning had struck her, she shouted, "Jacob! Jacob! Are you crazy! Get back! Have you not learned anything in your life?" This was the first time in their marriage that she had raised her voice to him. He smiled. Jacob straightened himself, flinging the broom at the ball in one final gesture to help the children. The cheerful sounds of the children rose from the alleyway as the ball fell to the ground followed by the broom a few seconds later. Jacob was very happy to hear their joyful voices.

"Promise me that you'll never do this again!" Sarah insisted.

"The probability of the ball landing again in that exact location is virtually zero, so you have nothing to worry about," he answered in his typical scientifically ambiguous manner.

On the first Sunday in June, Jacob ventured into the streets. In a relaxed frame of mind and armed with a handful of lollypops in his pocket, he walked along the surprisingly busy streets of the multicultural Cote des Neiges neighbourhood. Now and again, he stepped aside to allow hurrying young people to pass him. The city was diffidently awakening from a long stupor. Men and women were agog with joy at their rediscovered freedom.

On Cote des Neiges, a spring breeze herded last autumn's leaves across a front yard. Indian women in their native dresses, Sikh men with their helical turbans, and Haitians mingled with French Canadians adorned with tattoos and earrings. Jacob sauntered among them on his way to Beaver Lake. He thought, 'In this garden of humanity, God had planted a Jew to love them and to guide them.'

The fresh air made Jacob feel hungry. The slight cramp in his bowels never failed to remind him of the hunger he had endured as a child. He bought a banana at an outdoor fruit market. Nearby, people held shopping bags and trundled carriages. He wondered if any of them ever thought about God or saw divinity in the world. Did they have a smell for the spiritual life the way they smelled the flowers hanging from hooks along the outside the makeshift market wall? And, truthfully, he wondered if he was blessed with that special awareness? Had his futile attempts to reveal Godliness by means of a computer algorithm made him a better person than these simple, hard working people out on a Sunday stroll?

Across the fruit market, the once thriving Macdonald's Cafe had reopened but it was deserted. People wanted to be outdoors. Close by, a dog in the back seat of a car with its muzzle tightly bandaged by masking tape, stuck its head out the half open window, whimpering. The owner was standing at a Lotto kiosk, waiting in a long queue to buy tickets for the upcoming 45 million dollar first prize. Jacob unfastened the masking tape and patted the dog on the head with both hands. For a moment Jacob stood there, wondering what the true motives of his algorithm had been. Perhaps the search for the Messiah had been no more than a diversion in order to draw God into his corner of the rink and there to demand comprehensible explanations for the atrocities committed throughout human history.

As Jacob continued toward the mountain, he encountered a teacher walking backwards in front of a group of 10 three and four year old children secured by a rope like huskies tied to a sled. The children looked well fed. That made Jacob very happy even as he remembered a photo taken in the Warsaw Ghetto; a little boy dressed in short pants lying on a street, suckling cobblestones as men and women hurried past.

At a clothing drop-off container, a homeless man stood barefoot peeing. Jacob walked past the cemetery, then turned left off Cote des Neiges and began his ascent to Beaver Lake. A gaggle of teenagers stood on the crest of a hill smoking pot, the smell wafting across the roadway. Nearby, an injured young squirrel, its furry tail stripped off, was searching for food, moving slowly. Jacob followed it to a tree, lost in thought. 'Jacob, Jacob, are you really that squirrel?' he asked himself, then moved on.

He stopped at a boulder to his right, recalling the first winter of his marriage when Sarah and he would go for long walks here. The young couple was holding hands. Away from the eyes of the community, it was safe to do so. Sarah wore a short blue jacket. She was young and beautiful and Jacob had a thick red beard. They had stopped at an outcropping of rock on the side of the road covered in a curtain of vertically corrugated blue ice. Fifty years had flown by, yet the boulder was exactly as Jacob had seen it then.

Standing on the footpath circling Beaver Lake, Jacob saw an obese, white-haired woman in a white pant suit calling to a duck dawdling near the shoreline, "Here, cootchy cootchy," as she tossed breadcrumbs into the lake.

Close by, an Asian family had spread a blanket over the grass and was enjoying a picnic in the shade of trees. A young woman played a flute and a toddler was testing her father's patience with persistent forays to the lake. Jacob tried to distract the toddler by waving a lollypop. The child saw it and ran toward Jacob, her father behind her. Jacob handed the lollypop to the father as the child jumped excitedly up and down. After she calmed down, Jacob thought, 'Every incident and every thought, even a whim, has the potential to become the hypothesis of a living theorem. In the gray soil of the brain, concepts sprout forth like plants and in due course bear fruit.' Then he conjured up a new theorem, proved not by logic but by experience: *A green lollypop and a smile are sufficient conditions for peace in the world.*

From the observatory on Mount Royal, Jacob viewed the skyline of downtown and the buildings of his university where he had worked more than five decades. He had delivered thousands of lectures and graded countless examination papers - hopefully - with compassion. He had toiled on mathematical problems, but also had time to deliberate on Godliness.

In the late afternoon sunlight, Jacob stood on a boulder at the highest point of the park. He drew the *shofar* out of his breast pocket and blew it with all his strength, resolved to herald the good tidings of the Messiah's imminent appearance. The staccato sounds drifted through the surrounding hills and from there to the city center far below, echoing on and on.

ACKNOWLEDGMENTS

ABOUT THE AUTHOR

Abraham Boyarsky is the author of *Schreiber*, for which he was awarded the Canadian League of Poets' Gerald Lambert Memorial Award; *The Number Hall*, winner of the Toronto Jewish Literary Prize; *A Gift of Rags*; *The Ratcatcher*; *The Chassidic Trauma Unit*; *Through Shadows Slow* which won first prize for fiction in the 2020 Canadian Jewish Literary Awards; a number of short stories including the collection *A Pyramid of Time*; and the graduate mathematical textbook, *Laws of Chaos* with Prof. Pawel Gora. He lives in Montreal where he is a full professor emeritus of mathematics at Concordia University.

ALSO BY ABRAHAM BOYARSKY

Schreiber

The Number Hall

A Gift of Rags

The Ratcatcher

The Chassidic Trauma Unit

Through Shadows Slow

A Pyramid of Time

Laws of Chaos

BAYOU WOLF PRESS

Bayou Wolf Press is an independent publisher of quality fiction. If you enjoyed this book and would like to support us, please leave a review on Amazon, Goodreads, or wherever you review books. If you'd like to learn more about our press, sign up for our newsletter, and stay informed on upcoming books, please visit our website.

www.bayouwolfpress.com